The American Assembly, *Columbia University*

THE FARM AND THE CITY

RIVALS OR ALLIES?

Prentice-Hall, Inc., *Englewood Cliffs, New Jersey*

A SPECTRUM BOOK

Library of Congress Cataloging in Publication Data
Main entry under title:

THE FARM AND THE CITY.

(A Spectrum Book)
Background papers prepared for the fifty-eighth
American Assembly held at Arden House, Harriman, N. Y.,
from Apr. 10-13, 1980.
Includes index.
1. Land use—United States—Congresses. 2. Land
use, Rural—United States—Congresses. 3. Land use—
Congresses. I. American Assembly.
HD171.A15F37   333.73'13'0973   80-19998
ISBN 0-13-304980-9
ISBN 0-13-304972-8 (pbk.)

Editorial/production supervision by Betty Neville
Manufacturing buyer: Barbara A. Frick

10 9 8 7 6 5 4 3 2 1

PRENTICE-HALL INTERNATIONAL, INC. (*London*)
PRENTICE-HALL OF AUSTRALIA PTY. LIMITED (*Sydney*)
PRENTICE-HALL OF CANADA, LTD. (*Toronto*)
PRENTICE-HALL OF INDIA PRIVATE LIMITED (*New Delhi*)
PRENTICE-HALL OF JAPAN, INC. (*Tokyo*)
PRENTICE-HALL OF SOUTHEAST ASIA PTE. LTD. (*Singapore*)
WHITEHALL BOOKS LIMITED (*Wellington, New Zealand*)

Table of Contents

Preface   v

A. M. Woodruff and Charles R. Frink
Introduction   1

1 *A. M. Woodruff*
City Land and Farmland   11

2 *Brian J. L. Berry*
The Urban Problem   37

3 *Robert C. Weaver*
Coordinating Rural and Land Policy   60

4 *Charles R. Frink and James G. Horsfall*
The Farm Problem   73

5 *Robert G. Healy*
Landscape and Landowner:
Issues of Land Tenure in Rural America   90

6 *Frederick E. Smith*
The Environment   109

7 *C. Lowell Harriss*
Free Market Allocation of Land Resources:
(What the Free Market Can and
Cannot Do in Land Policy)   123

8   *Mark B. Lapping*
    Agricultural Land Retention:
    Responses, American and Foreign   144

    Index   179

    The American Assembly   183

Preface

In the years following World War II, the patterns of life and the patterns of land use in the United States changed dramatically. The population, which before the war had been distributed very largely in the cities and on the farms, moved in massive numbers to the suburbs. Single-family dwellings on small plots of land housed an increasing number of American households.

The center cities lost not only population, but also industries, as corporations followed their employees outside the city limits. With both these losses, they also lost substantial tax base and, hence, operating revenues.

The farms, which had been losing population as mechanization increased, accounted for less than 4 percent of the nation's population by the end of the 1970s. Most of those who left the farms, like those who left the cities, moved to suburbia or exurbia.

At the same time, prime farmland was voraciously consumed in real estate development for the suburbs. While statistics vary, it has been authoritatively estimated that by the end of the 1970s, prime farmland in the nation was down to about 380 million acres.

Much of the impulse which led to this change in the demographic patterns of the United States came from government actions responding to the growing political power of the suburbs. Road building, mortgage guarantees, subsidies, tax policies, and other forms of government influence accelerated the changes throughout the postwar period.

However, by the beginning of the 1980s, other influences were at work. The end of the baby boom, two-income families, gasoline shortages, inflation, and high mortgage rates were making the suburbs less attractive to a new generation of Americans. Moreover, there was increasing concern among many sectors of the population about the consequences which would flow from further loss of prime agricultural land. Several governmental jurisdictions have attempted, through various means, to preserve farmland.

In order to assess this changing situation and to recommend public policy, a group of distinguished Americans met at Arden House in Harriman, New York, from April 10 to 13, 1980, to participate in the Fifty-eighth American Assembly, which was entitled "The Farm and the City." Dr. A. M. Woodruff, President Emeritus of the University of Hartford, acted as director of this Assembly, and, under his editorial supervision, background papers were prepared for the use of the participants. These papers have been compiled into the present volume, which is published as a stimulus to further thinking about and discussion of this subject among informed and concerned citizens.

The opinions expressed in these chapters are those of the individual authors and not necessarily those of The American Assembly, which takes no stand on the topics it presents for public discussion, nor of The Lincoln Institute of Land Policy, which sponsored this Assembly.

> William H. Sullivan
> *President*
> The American Assembly

A. M. Woodruff and
Charles R. Frink

Introduction

When Pieter Stuyvesant, crusty director general of Nieuw Neder-land, strapped on his famous wooden leg and clumped down to the Bouwerie, he could, at the right time of year, see a parade of ships carrying food back to the "old country." The canny Dutch had established a colony based in New York harbor at the foot of the fabulously fertile Hudson Valley and just a hop, skip, and a jump across the East River from the even more fertile fields of Long Island.

When the Dutch were forced, in 1644, to turn the colony over to the British, Pieter Stuyvesant retired to his farm which was where St. Mark's Chapel stands today, in the heart of New York's financial district. Under urban regimes as various as the Dutch, British, Tammany Hall, and others, and comparable regimes in Albany, the real estate consistently followed two "iron laws." The first was that when strategic moments arrived and decisions had to be made, almost every

Educator, administrator, urban land economist, and advisor to government, A. M. Woodruff *is vice president of the Lincoln Foundation and president emeritus of the University of Hartford. Dr. Woodruff has been professor of urban land studies at the University of Pittsburgh, vice chairman of the Allegheny County Planning Commission, chairman of the National Capital Planning Commission (1960–62), and dean of the George Washington University School of Government. He has published extensively in the areas of business, land use, and economics.*

Charles R. Frink *is the chief soil chemist and vice director of the Connecticut Agricultural Experiment Station. Prior to joining that organization in 1960, he did research at Cornell University and the University of California at Berkeley. Dr. Frink is the author of numerous scientific writings, has received a number of awards, and belongs to several scientific organizations.*

owner put his land to whatever use would make him the most money. The second iron law was that, once committed, land remained in that use until the land by itself was worth more for other purposes than for any use the land could sustain in combination with the building on it.

Over the generations, farms in and near New York surrendered one by one to urban development. Pieter Stuyvesant could watch departing grain ships today, but he would note that the cargoes originate from Ohio and points west. The farms he knew have disappeared, following the first iron law, and, following the second, are most unlikely to come back.

Urban developers bid more for land than farmers could or would, and this process went on for 300 years after the Dutch turned the land over to the British. As developers took over mile after mile, the farmers moved farther out and kept on doing so, causing little concern over the "loss" of farmland until about the mid-1960s. Then concern began to be expressed that the country was nearing the end of the line. Farmers were tilling about 500 million acres and had no place left to go. City populations were still growing, and the need for adequate food was unabated. Several predictions based on the factors seen in the 1970s sounded ominous warnings.

The predictions of the 1970s lacked consistency, both of fact and interpretation, and a brief recital of details is essential to understanding the problem and the need for a firm data base. In 1975, a study by the National Academy of Sciences pointed out that three prior studies with target dates of 1950, 1960, and 1970–80 had underestimated population and overestimated food consumption per capita; all three failed to anticipate the American dietary shift from grains to meat which accompanied the prosperity of those years.

Two later predictions, one for the National Water Commission and one for the Commission on Population Growth and the American Future, corrected the inadequacies of the former studies but failed to anticipate the large food exports of the 1970s. These two studies relied heavily on returning diverted cropland to production. The Water Commission report estimated that fifty-eight million acres were reserved in the Soil Bank and that under the most severe conditions this reserve would be reduced to only 15.8 million by the year 2000. The Population Commission report estimated that the reserve in 1970 was ninety-six million acres, and their most severe projection reduced it to only thirty-one million acres by the year 2000.

Although in 1973 *all* land officially in retirement was released to production, by 1976 only thirty-seven million acres of reserved cropland had come under the plow, and the productivity of the released land was under that of land already being farmed. Obviously reserves were under expectation, both as to quantity and quality.

Estimates of the rate at which farmland was being converted to urban use ranged from one million acres per year to five million. The Soil Conservation Service estimated that urban uses were actually pre-empting three million acres per year, but leapfrogging and scatter-shot development were isolating another two million. The Soil and Water Resources Conservation Act (RCA) required predictions, and the most optimistic prediction from this source was one million lost acres per year.

The RCA study agreed with the National Academy of Sciences that production per acre might be tapering off, but projected a sizable population increase. It had three "scenarios," the first and most optimistic of which anticipated no serious food problem by the year 2030. The second scenario introduced some uncertainties. The third of these scenarios was more pessimistic and suggested that farm exports by 2030 would double. Yet other Department of Agriculture figures in 1979 showed that 45 percent of the export growth anticipated by 2030 had already taken place.

The RCA study noted that water tables were declining and energy costs increasing; and it estimated that deep well "mining" of water might be phased out by the year 2000. Nevertheless, this study assumed a major increase in the area of irrigated land. It agreed that water supplies were declining for cities, industry, mining, energy production, irrigation, livestock, and rural domestic use. Deep wells in some areas were producing water so "hard" that it had deposited salts on a number of fields, considerably impairing production.

The RCA study, instead of discussing the reserve cropland mentioned in the earlier reports, noted that 135 million acres of land not cultivated in 1979 could be tilled, although to do so would entail the loss of other products coming from this land, such as forest products. The uncultivated acreage consisted of 32.9 million acres of forest, 53.6 million acres of pasture, 41.5 million acres of range, and seven million acres devoted to other uses. The productivity of such lands would probably be noticeably under that of cropland actively farmed in 1980.

The RCA report also commented on the loss of soil through ero-

sion. Its conclusions were based on the "universal soil loss equation,"
which has been criticized on technical grounds. The equation, for
example, led to comments on the silting of Lake Waramaug in Connecticut and also the Housatonic River. Studies by W. A. Norvell and
others in 1979 and by Donald Aylor and Frink in 1980 indicated accumulation of nutrients in the lake but found no silting of either the
lake or the river.

Each of the three RCA scenarios was an extrapolation from about
the same data but involved different assumptions as to trends. Scenario
number three, the most pessimistic, said:

> All readily available cropland is exhausted. Some twenty-four million acres
> of potential cropland would also be pressed into production. Conversion of
> wetlands into cropland would increase significantly as would the total value
> of resources used in production. Irrigated acreage would roughly double its
> level, when exports are projected at 1975–1979 levels. Significant increases
> in pesticide use and commodity prices could also be expected.

Some experts questioned the comment on pesticides; most are petrochemical based and require considerable energy to manufacture. Hence
they will be scarcer and more expensive. In addition, more and more
bugs are getting more and more resistant to more and more chemicals
and blithely return to chewing and chomping America's food crops.

The rather dismal jeremiad of RCA ended with the following:
"Pending results from the National Agricultural Lands Study, minimize the conversion of prime farmland consistent with the secretary
of agriculture's memorandum #1827 on land use policy."

These predictions show an uncertainty of data, and their interpretation is reminiscent of the problem of the half-blind man in a very
dark room trying to locate a black cat that was not there. But the
effort involved in making the predictions and the fortitude displayed
by those who circulated them are extremely commendable. Only by
exposure to public comment and criticism could they be rectified to
the point of near consensus.

Farm and Urban Bidding for Farmland

Urban uses have a big margin wherewith to outbid farming for
land and water. Good cropland in the 1920s was worth about $250
per acre, which is about one-half cent per square foot. Good house
lots on the edge of medium-size cities were quoted up and down from
about forty dollars per front foot or four dollars per square foot. The
margin was about 80 to 1. Either figure could be adjusted in either

direction and the advantage remained formidable. Figures changed between 1920 and the 1970s; the margin shrank, but urban use retained a large advantage. Good farmland in the later years was quoted up and down from about $1,000 per acre; good outer-suburban house lots were quoted at so much per acre, at all sorts of figures in different places, but $15,000 per acre was reasonably conservative. This left a margin of 15 to 1, much less than the 1920 ratio, but still enough to command the market. With events left to unfold themselves, urban uses could win every time, farm by farm and mile by mile. But pressure for further urban expansion seemed to be ebbing in the late 1970s, and pressure to preserve farmland was building up.

American cities grew fast in the post-World War II years, and at the same time new factors were added to the rural/urban equation. International food resources were far short of total international requirements. America, in no position to import food in view of its international energy balance, could hardly look beyond its borders for additional rations. As American cities expanded horizontally, they ate up farmland in the process, and no one could predict when the process would stabilize. If the 1945 to 1975 trends were to continue, so much farmland might be urbanized that only the rich could eat well. Facing such a specter, even a vaguely perceived specter, many felt that the third RCA scenario was the one on which plans should be based and that government should intervene to protect farmland from much further encroachment.

Changes in Urban Patterns

In the late 1970s the rate of urban horizontal expansion seemed to slow down, although conclusive figures were hard to find. The minor gasoline crisis of 1974 scared a lot of city people, and the major crisis of 1979 scared them far more. As gasoline rose toward two dollars per gallon, the price forced rethinking of priorities within the family budget.

Urban development of the period from 1945 to 1975 had left big holes inside the limits of standard statistical metropolitan areas. Leapfrogging and scatter-shot development had by-passed tracts. Large holes were left "far inside" by the outward relocation of industry. Finally the mass migration of displaced farm workers, who were initially ill-adapted to northern urban living and unprepared for urban jobs, had a major impact on urban evolution, affecting their own lives and those of everybody else.

Like all refugees, they arrived without funds and, like all the very poor, they had to live in the cheapest housing they could find. This was in the former tenements of earlier immigrants and the remains of once fine homes left behind as the former occupants "moved out." The migrants poured into these dwellings. In most cases the condition of the houses was not very good to begin with, and the extreme overcrowding by the migrants led to rapid deterioration. While taxes rose and mortgage payments did not diminish, low rent, aggravated by compassionately inspired legislation like rent control, kept landlords' incomes down so much that they could not both make mandatory payments and keep up with repairs. Property was abandoned wholesale and was brutally vandalized. New York City had about 8,500 apartment houses vacant and more or less vandalized; many had been virtually gutted by the vandals, but many of the surviving structures were still sound enough to warrant reconstruction.

New York also had nearly 3,000 vacant duplexes and about 1,800 vacant single homes. It also had nearly 950 lots left vacant by residential demolitions of buildings too far gone to repair. Altogether there were 588 vacant industrial buildings in the five boroughs and thirty significant plots where industrial buildings had once stood.

If the residential "vacancies" could be fixed up sufficiently to permit occupancy, and if the average capacity was twelve families per apartment house with five members per family, then the idle apartment areas could rehouse most of the population of Flint, Michigan, slightly more than 520,000 persons. Even with smaller family units, the numbers would be massive. The duplexes, at two families per house and five-person families, could have housed 30,000 more, and the vacant singles, about 10,000. This adds up to the considerable total of 560,000 people. Even at half this density the numbers remain impressive.

The 950 vacant parcels (547 apartment sites, 200 duplex sites, and 203 singles) could have been rebuilt to house 36,000 at the density assumptions of the previous paragraph. This still would have left 588 empty industrial buildings and thirty vacant industrial sites. In some cities old factories were being made over into apartments, not the best housing in the world, but still far better than none at all. To summarize the foregoing in one brief sentence: New York had holes enough to house a medium-size city, and if all these people had to live on half-acre suburban lots, about 70,000 acres of erstwhile farmland would have to be converted. These are, of course, extreme suppositions; families were getting smaller, and while far-out suburban zoning was tending to larger and larger lots, ways could probably have been found to house people more compactly than on sprawled half-acre lots that many people could not afford.

Other significant urban developments were the wholesale conversion of in-town apartments from rental to condominium, as landlords were harder and harder pressed to meet payments. The tax advantages of condominium ownership are described elsewhere in these chapters, but many former tenants balked at the cash "front-end" payments involved for them in the changeover. Others simply could not make the payments. This left a lot of impoverished and semi-impoverished people, many very elderly, in a major housing crunch for which no clear outcome was in sight.

At the same time, the "Georgetown" syndrome was appearing in many cities. In the 1940s and 1950s, Georgetown, for generations a congested, poor neighborhood, was "discovered" by the upper-middle class which swarmed in, outbid the former occupants, and pushed them out, acquiring fine old structures to rebuild in all their eighteenth century glory. The convenience of close-in living appealed to those weary of long automobile trips to the city of Washington.

Finally, a revolution in urban transportation was quietly taking place. Several cities had built rail systems, notably Washington, D.C. and San Francisco. The Washington system, mentioned elsewhere, was opposed by many, including the automobile and highway lobbies, but it had by 1980 begun to fulfill the planners' expectations. People began to cluster near transit stops, and the same thing seemed destined to happen outside the city as lines were extended into the suburbs. The reconversion or remodeling of old houses was becoming something of a national trend, and most cities were "doing something" about mass transportation.

The combination of all these pressures seemed to adumbrate a move back to the convenience of city living. Smaller families, families with no children, and families in which both the husband and wife pursued careers were especially attracted to the convenience and elimination of commuting time which in many suburbs amounted to more than 25 percent of the time spent actually "on the job." The move "back to town," beginning as a trickle, began in 1980 to look like it might be the start of a flood.

Urban planners argued that cities should grow upward and slow the outward sprawl, partly to conserve energy, partly to fill up the urban holes, and partly to make better use of well-established downtown cores. Many offices had followed the factories out to suburbia, and this had led to suburban clusters. But in the last years of the decade of the 1970s, construction of downtown office buildings resumed in many cities. As new jobs developed in the core, the incentives to urban living accelerated. The demand for more land on the outskirts for urban expansion seemed in 1980 to be slowing. If this proves more than a minor interruption of the 1945 to 1975 trend, it could usher in a fairly long period of relative farm/urban stability. The farm and urban uses were less antagonistic to each other in 1980 than at any time since World War II. But developers with a vested interest in building houses on detached suburban lots cried out lustily against any interruption in the market for land, and, pointing to the numbers of the ill-housed, argued that the demand for new single houses was enormous.

Alternatives to Traditional Farming

Through the 1970s, discussion was heard of various ways to feed people without much use of land, as an offset to land disappearing under the concrete and asphalt of suburbia. Hydroponics as a way to grow food still required room to put the plant roots. The sea proved to be a less abundant food source than had been assumed by some in the 1970s. Some varieties of marine life had been overfished almost to the point of extinction. Enough miracles had emerged from America's laboratories to discourage exaggerated pessimism; ways might be found to draw far more food from the sea and to grow fish commercially in ponds, but they did not hold quite the bright promise in 1980 that they might have.

The decline in the effectiveness of pesticides was already noted, as was the fact that additional inputs of fertilizers used in the 1970s seemed to have passed the point of diminishing returns. Costs of additional fertilizer inputs were not justified by increased yields. Organic food raising appealed to the "health food" set, but as an alternative to increasingly scarce fertilizer, it was hardly a national alternative. Animal manure was hard to ship and unpleasant to handle, and fastidious urbanites tended to recoil from its pungent aroma. There was nowhere nearly enough animal manure for all needs. Leaf compost, which proved to be an effective substitute in some controlled experiments in the Northeast, was also less than a national alternative. The prime middle western cropland was short of trees.

Environmental laws discouraged the draining of wetlands and the filling of ponds. Some desert land could have been made to blossom, both in America and all over the world, if there had been water enough. But available sources were approaching depletion, and desalination of sea water had proved economically practical only in countries with a great water shortage like Israel. Better methods of desalination using solar energy in place of fossil fuels could make a big difference, but this alternative was not in sight in 1980.

America could, through major changes in diet and living habits, cut down on the space needed for food. The American diet, heavily dependent on meat, could have been shifted to one using balanced vegetable protein. The vegetable proteins require far less acreage than the feed grains that must be shoveled through meat animals in order to place beefsteaks and pork chops on the American dining table. The Chinese for centuries have used meat more as a garnish and less as a staple than Americans, and on Taiwan the average consumption is about 2,050 daily calories with abundant protein. Yet, while Chinese cooking appealed greatly to American gourmets, there was no national trend away from the beloved beefsteak and hamburger.

Yearly, Americans feed something like five billion dollars worth of pet food to their dogs and cats. This practice could be discontinued and this food stream diverted to human mouths, but American pet owners show no visible sign of making such a change. In an extremity, Americans could eat their pets, but the very idea is offensive to many, although dogs and cats are part of *haute cuisine* in some places.

Purpose of This Book

Amid the conflicting viewpoints, this book does not suggest pat answers but rather seeks to clarify serious questions. Well-formulated questions may be hard to answer, but they are far more manageable than questions which are unclear. The authors of these chapters addressed themselves to questions like the following:

1. How much farmland had been irretrievably converted to urban use from 1945 to 1980, and what was the 1979 rate of conversion?
2. What is the future capacity of America's farms? How real are the threats of diminishing productivity?
3. Recapitulating the arguments, how persuasive is the case that farmland should be preserved for farming?
4. If a decision is made to preserve farms, what is the best way to do it?
5. Several American states and a number of foreign countries have tried to preserve farmland. Why did they decide to try? How well did their systems work? And what were the major obstacles?

The foreign countries that preserved farmland are much smaller than the United States and felt threatened by the menacing grumbles of large and powerful neighbors. The Dutch, who fought the sea for centuries, "polderized" new land after World War II, and the loss of the Dutch East Indies closed the escape hatch for surplus population at home. Israel, outnumbered by its neighbors, followed a cautious food policy from the start. Taiwan, threatened by the mainland, had similar reasons to preserve farmland.

A decision to preserve farmland raises the additional question of whether the shortfall between the landowners' expectations and realizations justifies compensation. Unfulfilled expectations are not new; past expectations die hard. Urban homeowners in the late 1920s thought their property was worth a lot of money, only to have the rug pulled out from under them by the depression; many held on until they "got their price" in the 1940s or 1950s, though it was then paid in relatively undignified dollars. Farmers developed expectations and many kept on farming only because they thought their hopes would eventually be realized.

The courts long since decided that businesses like motels, gas stations, and others with profitable highway locations could not be paid for loss of business when a new superhighway paralleled the old route and syphoned off most of the traffic. Many were wiped out; such situations were not uncommon in the period of active superhighway construction. The same holding applied to property that lost value through change of zoning. These cases were less frequent since zoning was done locally and those aggrieved could howl right into the ears of those making the decisions. Up-zoning (to more intense use) was common, but down-zoning (to less intense use) was rare.

Zoning of fringe land exclusively for farming would have frustrated the expectations of many a farmer and speculator. The lamentations of the frustrated combined with the persuasive rhetoric of the developers made local "down-zoning" an unlikely solution.

This raised the question of whether good policy required payment of some sort of compensation, notwithstanding the lack of legal necessity to do so. It led to the final question this book tries to raise.

   6. Should owners frustrated by zoning be paid? Should directive programs like selective mortgage financing be used instead of zoning to divert development out of the fringe? If payment is in order, then how much and how should it be financed?

A. M. Woodruff

1

City Land and Farmland

Land Switched to Urban Use

Throughout the United States, especially between the end of World War II and the mid-1970s, many acres once green with field crops were switched to other uses. The rate and magnitude of the switch has been considerably debated. The lowest figure commonly cited was one million acres per year. By 1975 this had been generally raised to two million; then in 1977 the Soil Conservation Service raised the estimate to three million and added that for every acre actually urbanized, one more was isolated by leapfrogging. Chapter 4 has more to say on the subject of loss; on the basis of any estimate, it was considerable.

Land switched from field crops to other agriculture-related uses can be switched back under proper market conditions. Land switched to urban uses is generally switched "forever." Forever is a long time, but land thus switched is unlikely to return to farming for centuries, if even then. In the 1930s the nation had started programs to limit farming and reduce crops and in 1956 had set up soil banks subsidizing farmers to curtail production by putting land out of crops and into the "bank." No certain figures show how much land was banked, but when in 1973 and 1974 all privately held land was released for production, only thirty-seven million reserved acres returned to production out of a conservative estimate of sixty million. The productivity of the land returned was far under expectation. America had little reserve of really prime farmland.

Uncertainty of definition plagues the question of urbanization. Is land urbanized when:

1. the land is divided into recorded plots?
2. the land is incorporated within the boundaries of cities which are absorbing "outside" land or within boundaries ascribed to standard metropolitan statistical areas?
3. the land is actually built upon or paved over to make highways or parking lots?

The third criterion is clear enough but extraordinarily hard to quantify. The first two permit easy quantification, but some land would be "urban" by criterion one or two, but not by number three. Farming and gardening linger in the urban fringe on some land which is "urban" by the first two criteria.

Active gardening extends right into the heart of cities. Urban community gardens and suburban gardens, when properly tended, produce impressive yields; small farms of twenty to fifty acres flourish in the suburbs. The big gardener and/or small farmer need places to buy the proper machinery to work the land and places to get it fixed when it breaks. The small machine business in rural areas is lacking but, responding to demand, flourished near the cities in the period from about 1960 to 1980. Small farmers also need better ways to sell crops than at their own roadside stands, and this need at the time of this writing remains largely unmet. Such cultivated land is "urban" by some systems of classification, but it is not "lost" to food production. The food-producing capacity of the urban fringe is discussed in more detail later in this chapter.

Urban fringes with large lots and considerable unbuilt areas support more wildlife per acre than strictly farming areas. Brush and untended groves invite browsing animals from rabbits to deer. Predators which normally follow the browsers are not welcome in decorous suburbs, and the browsers pose a real threat to suburban gardens.

Since 95 percent of all Americans in the early 1980s live in cities or large towns and only 5 percent live on farms, the numbers suggest that the problem be approached in light of the best interests of the urban population. The urban people obviously need enough to eat. As America continues to mature and become more densely populated, choices have to be made that were not required when a small population and an expanding frontier allowed profligate land use.

The urgency of land preservation depends partly on how "farmland" is defined; the extent that American agriculture could expand production or, conversely, the extent to which it faces possible shrinkage; the need for foreign exchange to buy imported oil; and imponderable changes in the standard of living.

A diet with more vegetable protein and less meat would reduce demands on middle western farmland. Solution of the energy problem within America's own boundaries would relieve one economic need to export food to pay for imports, leaving "compassion" exports to help ease world hunger. A gross error in either direction in population forecasts or any other assumption could change all other estimates. Considering the past record of population forecasts, gross errors are not a remote possibility.

A "best case" forecast assumes fulfillment of all euphoric predictions and shows little cause for immediate alarm. A "worst case" forecast assumes the accuracy of all the pessimistic predictions and is truly alarming. Because of permanence of shifts to urban use and despite the remaining food potential of urban areas, the worst case develops a grim argument to preserve farmland. Land preserved for farms can later be urbanized, so this course of action would leave later options. Land once urbanized would probably be urbanized forever and leave no options.

Some city planners currently argue that preservation of farmland on the urban fringe would encourage cities to develop more compactly, and that this would, in turn, improve the quality of urban life. Compact cities are more economical in many ways; utilities per connected family are considerably cheaper in dense areas where the number of users per mile of service is high. City streets are cheaper per user family than rural roads for the same reason. Heating costs for families in apartment houses are substantially lower than equivalent costs for detached houses, and the differential goes up as the exposure of the house increases. Summer air-conditioning costs work the other way; the detached house is easier to cool than a property in a dense part of a city. Private commuting costs go down as density goes up, and the public costs per rider of subsidized commuting facilities are lower in dense cities. Finally, this group of urban planners points out that enough land had been urbanized by 1980 to accommodate any predictable growth without urbanizing much, if any, more, provided the land already urbanized could be better utilized.

Other planners argue that free market forces should be left alone. As long as land is worth more for house lots than for farming, the market has spoken and land should be left to find its own most profitable use. The weakness of this argument is the permanence of an unwarranted urban shift compared to the flexibility which a judicious policy of land preservation would allow future generations.

Population Fluctuations

Land was switched to urban uses as cities grew. Urban growth rates in turn depended on the raw numbers of population. Lateral expansion followed the rate of net new family formation, which followed the birth rate of about twenty years before. Since birth rates were high in America during the early 1960s, the rate of new family formation is expected to remain fairly high in the early 1980s. The demography is discussed in detail in Chapter 2.

Population predicting is a less exact science than weather forecasting. Predictions are made by extrapolating from the past, and fluctuations, recent and long past, have been considerable. One of history's classic fluctuations occurred long ago after the plague of 1350 struck down about half the population of western Europe. Surviving records indicate that it took less than a century of fast breeding to rebuild numbers; and after they had been rebuilt, the growth rate subsided to the maintenance level.

Population growth rates almost always rise after a disastrous war or other catastrophe. Napoleon's wars decimated western Europe and a major baby boom followed the establishment of peace in 1815. Each successive generation of boom babies produced a boomlet twenty or so years later, and these waves could be traced up to World War I. America had a similar baby boom after World War II which subsided in due course, with a sequential boomlet in the late 1970s.

Humanity responded, when threatened, to some deep-seated instinct for survival and the need to upbuild numbers. This plus other uncertainties makes future growth rates hard to predict, and if America has any large increase, it will need land to feed the extra mouths.

Three Different Points of View on the Land Shortage

NATIONAL

The United States has had recurrent crop surpluses since the end of World War I. A very serious glut occurred at the end of the 1920s and during the early 1930s and led to restrictive government programs like those of the Agricultural Adjustment Administration and the Soil Conservation Service. A new surplus loomed on the horizon in 1980 when the President declared an embargo on grain shipments to Russia.

Productivity per acre increased so much between the end of the war and the mid-1970s that many observers thought it could be increased still further, as if by magic, whenever the need arose. Chapter 4 of this book concludes otherwise. Nevertheless, the surpluses, the productivity increases, and euphoric dreams of further increases were woven into the argument that America, from a national standpoint, still had excess capacity and could complacently switch considerably more farmland to urban use.

The spectacular increases in grain yields came from the introduction of hybrid crops and the application of high-potency agrochemicals —fertilizers, pesticides, and herbicides. Few hybrids were developed in the 1960s and 1970s. The potent agrochemicals require energy in the making, and most are synthesized from a petrochemical base. As the price of oil increased, once economical compounds became costly by 1980.

The ability to improve yields further by still more intensive use of fertilizer is questioned by many farm experts who feel that its use was optimized in the 1970s. They feel that pesticides and herbicides are backfiring as successive generations of bugs become resistant to them and as humans are threatened by the compounds and their residues.

Erosion caused by overfarming is a further threat, as is salination of land caused by heavy irrigation with "hard" water, i.e., water carrying heavy burdens of salts which build up in the soil. Most well water is hard, as is some river water. Rain water, by contrast, is "soft," and countries with heavy seasonal rainfall which can impound large reservoirs of rain water can irrigate more effectively than western farmers in the United States, many of whom depend on wells.

Chapter 5 discusses changes in the pattern of landownership, some of which add to the threat of lessened productivity per acre. Larger and more expensive farm machinery requiring more land to support it and saving only man-hours, not land, gradually began to eliminate the "family farm." This movement began in the early postwar years and accelerated in the decade from 1970 to 1980. The family farm is by no means extinct but is becoming something of an endangered species. The ownership ratio was fairly high during the height of the family farm period, and owners, especially fairly young ones who anticipated many years of active farming, tended to husband the fertility of their property.

"Overfarming" was a problem in the early 1930s when many farmers faced mortgage obligations undertaken right after World War I. Inflexible mortgage payments had to be met out of income which had shrunk in the price collapse of the late 1920s and early 1930s. Individual farmers needed all the cash they could get for their payments and ignored the then conservative practices of crop rotation. Their big crops added to the price-depressing surplus, and the action to which each individual was impelled made their collective situation worse.

National policy reversed from "strictly hands off" to limited public control. One argument in favor of control was the steady depletion of farmland fertility through overfarming as wind and water erosion took a huge toll. The dust storms of the period were legendary, and little rivulets washed out gullies, some big enough to hide a trailer truck. The public response, endorsed by both major parties in the election of 1932, was a plan to limit crop acreage together with various programs to halt erosion and protect farm integrity. Such programs followed quickly in the mid-1930s.

The ownership changes of the 1970s brought considerable land into the hands of corporations and foreign nationals who bought it primarily to participate in anticipated capital gains. Whereas the "personal" landlords generally were retired farmers or other locals who knew the land, visited it often, and saw to the tillage, the new landlords lived at great distances, and their interests were in profit rather than long-term preservation. Especially if the land were held for eventual conversion to urban use, fertility came second. The tenants, under these circumstances, knowing that they could not attain ownership, tended also to push for short-run profit. The combination of these factors seem to be reviving some of the hazards which the 1930 programs were designed to forfend, especially wind and water erosion resulting from overfarming.

The results of future research cannot be discounted. New miracles may well emerge from the laboratories, but "when?" and "what?" remain unanswered questions. An uncertain research future is a slender straw to grasp as known means of sustaining food production disappear. The rising cost of petroenergy, the probability that America might have passed its productive peak during the 1970s, and the uncertainty of future population growth add weight to the contraargument that further switches of farmland to urban uses should be most carefully scrutinized.

REGIONAL

Lingering national complacency is balanced by regional anxieties. The East Coast megalopolis extends from Portland, Maine, to Richmond, Virginia, and housed about 40,500,000 people in 1980 in thirty-two standard metropolitan statistical areas plus other areas adjacent to them. Altogether about 20 percent of the 1974 American population was in this megalopolis. That region, once overflowing with its own "milk and honey," produced only a fraction of its own food by the late 1970s and depended largely on shipments from places like California and Florida. These people face higher costs as increased diesel oil prices make trucking more expensive, and they face actual shortages whenever labor troubles erupt in the fields of their supply areas or when strikes shut down shipping facilities.

The world was clearly in a mess at the end of the 1970s. Several Third World countries, such as Iran, Nicaragua, and El Salvador, had experienced wild revolutions. America's international prestige was at a low ebb in the late 1970s, while Russia charged off on the expansionist course to which the tsars had been once committed. The Third World countries were once again choosing sides for what was essentially a new round of international gamesmanship under new and different rules. Under such circumstances, a major international war is always quite possible, and no longer can America anticipate geographical immunity. The nation's transportation network can be seriously disarranged beyond the hope of speedy repair, and a month without food shipments will leave many on the East Coast hungry and a few in deep distress. America is peculiarly vulnerable to transportation stoppages, since consumers stockpile little food in their homes and put great confidence in the distribution system which brings food close to most doorsteps via the ubiquitous supermarket.

The same factors, in varying degrees, apply to other megalopolises, such as the one south of the Great Lakes. This reasoning plus, incidentally, the greater palatability of local produce over varieties adaptable to long-distance shipping, became the fabric of the regional argument that local farmland should be carefully preserved in and near areas of great population concentration.

GLOBAL

The third point of view is global and involves other factors. World population in the late 1970s was about 4.5 billion and growing fast. The medical revolution of the nineteenth and twentieth centuries had reduced infant mortality; but breeding habits, developed over the centuries when large families were needed to assure that some children would survive, lingered long after the need disappeared. Population grew fast, and about one person in four was chronically hungry in the late 1970s, and about one in ten scraped along barely above the subsistence level. The number of the barely subsisting equaled more than twice the whole American population.

Switching of farmland anywhere, with so many people hungry, was widely considered an immoral act of wanton disregard for human misery. But much human misery has been visited on local populations by preemptive rulers romping about on wars of aggression and personal aggrandizement. Several areas, including the Mekong Valley in Cambodia and Vietnam, could produce bumper surpluses of rice, if the unruly dogs of war could be chained up in their kennels. The obligation of the United States to help feed millions, who could feed themselves under different political circumstances, raises a moral question which this book does not try to answer.

From a pragmatic point of view, unstable countries, whose people have grumbling, painfully empty bellies, pose a vague but constant threat to countries like the United States with large food-producing resources. Sir William Blackstone, the legal commentator, argued in the eighteenth century that the colonizing activities of the Europeans were justified because Europeans would use the resources more effectively than the locals. Third World revolutionaries use this argument in reverse to show what 800 million impoverished people could do with America, which to them seems sparsely populated.

American foreign trade in the 1970s was financially disastrous. The national economy had come to depend on imported oil and then, when the price rose, had real trouble paying for it. Presumably technology would rectify the energy situation in due course, but meanwhile foreign exchange was needed badly. American industrial productivity seemed to be falling while that of the Third World was rising, and America again looked to its bountiful farms to produce the foreign exchange it needed.

SUMMARY

The various arguments can be summarized as follows:

1. America cannot count with assurance on increasing farm productivity but might well face a decline.
2. From the standpoint of the country as a whole, the problem is serious, even though no immediate crisis overshadows the country.
3. Some regional shortages, especially in the East and Northeast, develop whenever the transportation network is disrupted. They have been neither serious nor longlasting, but the potential for trouble is ever present. Regions of heavy concentration of consumers and small local production have real cause for anxiety.
4. America needs farm surpluses to maintain its balance of international payments. A temporary interruption like the embargo of 1980 does not alter this basic fact.
5. From a global point of view, the world needs more food.

The weight of the argument seems to indicate that America should not be lulled into complacency by short-run surpluses. A long-run look suggests that the need to preserve farmland, if not conclusively proved, is at least strongly indicated.

The Expansion of Cities—The Question of Densities

Crude population was just one factor among many that shaped cities as they grew. People could live either in compact cities or in "sprawled" urban areas. In 1980 New York was the most compact city in America and one of the most compact in the world, with a density of 26,000 people per square mile. Other than New York, East Coast cities of 100,000 or more people *within* their borders averaged, in 1970, about 10,000 per square mile. Cities of 100,000 or more people south of the Great Lakes average 6,200 per square mile; California cities averaged 4,500 per square mile, and Texas cities, 2,700 per square mile. Elsewhere in the world, urban densities were higher than most American averages. Hong Kong had a density of 11,200 and Singapore, 10,200. Kaohsiung on Taiwan had a density of 9,100 and Taipei, 7,500. Other Asian cities were as dense or more so. Large European cities were also dense in comparison to America.

American cities sprawled out from high-density cores into low-density suburbs and then into exurbs of still lower density. This pattern is rare outside America; in Asia the apartment houses marched in lockstep right up to the line of urban development, and rice fields lay beyond. American East Coast cities generally have geographically small cores, and the suburbs are politically independent. The apparent density of cities depends on this "balkanization." Fractured cities in the East seem to have higher density because they do not include lower densities outside. Several western and southern cities extended their municipal boundaries as they grew to include low-density areas from "outside."

A typical pattern of "balkanized" suburbs is found around Hartford, Connecticut, geographically one of the smallest cities in the East. Twenty-eight independent suburbs fill the rest of Hartford County. The city's urban density of 8,400 is more than twice that of the two densest suburbs, East and West Hartford, each of which has just over 3,000. Twenty of the suburbs have densities under 1,000. The lowest density is 123, the median is 559, and the mean is 907. The county has 758 square miles. Of these, 582 square miles (or more than 75 percent of the area) have less than 1,000 persons per square mile.

Urban switching by no means preempted the whole of the county, but creeping and leapfrogging development put quite a crimp in farming. The creeping process filled the towns of 3,000 density about full with single houses, and very little farmland remained. Beyond that, the leapfrog pattern was more pronounced than the creep. The developer seeking a piece of land did not care much whether it adjoined other developments; hence, he prowled the countryside until he found a tract that suited him. It needed access to a good town road, and, since the mid-1970s, it had to pass a rigid percolation test to establish its capacity to absorb septic tank effluent. The availability of town water helped greatly but was not entirely essential if wells could be obtained at a reasonable depth.

The land search often took the developer across a considerable swatch of farmland before he found a tract to suit him. Hence, the term "leapfrogging." He leapt over farmland and established a new urban area beyond. The by-passed farmland lost little usefulness at first, but as urban population closed in on it, the inconvenience of farming became increasingly evident. Urban-type neighbors objected to slow-paced machines moving on "their" roads from farm parcel to farm parcel, and they became profanely eloquent over the movement of manure. Chapter 4 discusses these problems more fully.

The chief point of mentioning the large area of low density in the outer suburbs is that a very large addition could be made to the population of Hartford County without slopping over county (or standard metropolitan statistical area) lines. Assuming that East and West Hartford set a norm of 3,000 per square mile, the 582 square miles at lower density (averaging in the late 1970s about 600 people per square mile and accommodating a total of about 350,000 people) could accommodate about 1,750,000. This would add 1,400,000 and about triple Hartford County's present population of about 825,000.

This very rough calculation is no firm guide to the absorptive capacity of other suburban districts, but it illustrates the generalization that areas already substantially urbanized could handle large additions without switching much farmland. And this could be done at single-family densities on lots of about half an acre. If half the urban population lives in apartments, as does most of the urban population of Europe, and the other half in singles, with an average density of about 8,500, then Hartford County could accommodate as many as 6,500,000 people.

Such illustrative calculations do not prove a lack of need for additional urban land but suggest that considerable growth would be possible without going beyond 1980 standard metropolitan statistical area boundaries. They also suggest that since the farm efficiency is higher if farm and urban development are somewhat separated, the zoning power could well be used to keep the zones apart, just as the power was traditionally used to segregate manufacturing districts.

The Forces That Shape Cities

CONSUMER PREFERENCES

In the United States most land use decisions are made by private citizens, both developers and consumers of housing. The consumers, by their collective decisions both in renting and buying, create local areas of greater or less demand. Some places are perceived as "good neighborhoods," with "good" school systems, and usually with an absence of many very poor people. These perceptions are subjective and depend on how people regard their neighbors. Status conscious "taste setters" seek "fashionable" addresses. Those who follow after try to emulate on a more modest scale the housing mores of the leaders. The Tenth Commandment which says, "Thou shall not covet thy neighbor's house," is almost universally honored in the breach.

THE MORTGAGE MARKET

Consumer preferences did not develop in a vacuum; most home purchases depended on mortgages. The mortgage lenders were a relatively small group and competed fiercely with each other, but they nevertheless followed a well-developed herd instinct. If lenders generally favored certain localities and avoided others, then individual lenders tended to do likewise. The price of nonconformity could be very high. If something went wrong with a loan in a neighborhood where "everybody else" was lending, then the particular lender in question was not immediately assumed to have violated the "prudent man" rule. If, on the other hand, a lender went pioneering and some loans went sour, the opposite assumption arose. One bank went to considerable lengths to make loans to try to bolster up the less affluent parts of a sagging city, but suffered many delinquencies and some foreclosures. Down on their luckless heads came the wrath of the bank examiners who forced the president to resign and threatened to close the bank.

Since 1933, the federal government has strongly influenced the mortgage market. The Federal Housing Administration (FHA) was established in that year to insure high-percentage mortgages made by private lenders. A major objective was to stimulate a sluggish house-buying market, which in turn would put back to work a lot of workmen in the building trades.

The FHA in those days insured mortgages up to 90 percent of the appraised value of houses built under its scrutiny, but only 80 percent on existing houses. Appraisal protocols also favored new structures, and young couples looking for homes found they could do better in the suburbs than in town. "Conventional" (i.e., uninsured) mortgages were then usually for 66 percent. These percentages rose in the next quarter century, but the FHA long dominated the market with its high-percentage loans, and the conventional market followed. The movement to suburbia was fueled by the actions of the FHA. Whether the mortgage market led or followed the consumers was much debated, but the FHA influence was enormous.

After World War II, returning servicemen were entitled to Veterans Administration (VA) loans. These were guaranteed by the government and, like the FHAs, made by private lenders. The VA relied on broker appraisals, whereas FHA used staff appraisals, and for a while after the war VA appraisals were consistently higher. The VA also guaranteed loans on older houses, and many an owner of such a house finally "got his price" from a returned veteran. As the postwar boom petered out, some older houses were offered at "two prices," a higher price to veterans and a lower one to a buyer who had no VA entitlement.

Notwithstanding the flicker of life in the old house market, the bulk of the housing for millions of returning veterans was in suburbia. Once sleepy country towns awoke and became bustling suburbs. Supermarkets and shopping malls followed the people. Newcomers demanded urban services like water and sewers, and they sought and won public offices and gradually took the management of the town away from the older rurals, to the distress of the latter.

Apartment construction followed the same basic rules. Mortgages were required for virtually all projects, and the mortgage market had the same ambivalent "hen and egg" relationship with the consumers. After World War II, garden apartments were built in great numbers in the suburbs. These were two- and three-story (seldom higher) non-fireproof buildings without elevators, built in relatively spacious grounds, instead of the taller apartments with elevators built right to city lot lines. Garden apartments generally sought locations where building codes did not require fireproof construction, and this often meant politically independent suburbia.

Returning World War II veterans, demobilized in sudden numbers, needed housing quickly, and the FHA instituted its "608 program" which financed 90 percent of the *cost* of building apartment houses. Appraisal procedures let an efficient builder erect a unit without front-end load. This produced apartment houses at record rates, and most of them turned out to be long-run commercial successes. This shows how a mortgage program can influence a market.

The Veterans Administration gradually became less active as the very large numbers of servicemen from World War II used up their entitlement. The Federal National Mortgage Association (FNMA) joined the federal lending family in 1938. It was shortly nicknamed

"Fannie Mae"—every borrower's girlfriend. "Fannie Mae" assisted in the liquidation of distressed mortgages, operated in the secondary market, and was a lender of last resort. It made loans, usually FHA insured, to persons and in places the private market avoided, using funds obtained from government bonds. "Fannie Mae" became semi-independent in 1968 and is still in existence. In 1968 the Government National Mortgage Association (GNMA) was formed as part of the Department of Housing and Urban Development and was soon nicknamed "Ginny Mae." It is primarily a lender of last resort when the conventional market is not functioning.

In the late 1970s the mortgage market virtually dried up. Interest rates rose to the highest point in the twentieth century, and funds were scarce. The rate of new construction fell off, and with it both the leapfrogging and the urban crawling.

TAXES

The federal income tax came of age during World War II and thereafter exerted more and more influence on consumer behavior. This tax grossly favors homeownership over rental, since taxes and mortgage interest on a house can be deducted from the base on which income taxes are computed, while no such deductions are allowed to renters. Since homes could be bought in the suburbs with favorable mortgages, the income tax added force to the centrifugal movement.

Property taxes are more complex. Right after the war, suburban property taxes were almost uniformly lower than the urban equivalent, but new residents demanding urban amenities tended to increase governmental costs, and the suburbs gradually lost some, sometimes much, of their tax advantage.

The outward march of the working middle class was accompanied by a dramatic relocation of industry (which is discussed later) and was followed by the relocation of mercantile facilities in suburban shopping centers. These grew in size and elegance, and by the late 1970s became "shopping malls." As the cities gradually lost industry, commerce, and middle-class residents, the three workhorses which had carried the burden of their taxes, a new poverty population, migrating from the South to northern cities, needed a disproportionate amount of public service. Tax rates rose as the tax base shrank, and this boosted some city taxes toward the fiscal stratosphere, thus encouraging still further moves to suburbia.

Some American cities, sorely beset, economized in ways that still further aggravated the middle-class exodus. Police and fire protection were reduced; garbage was collected less frequently; and streets were left unrepaired. These factors, especially the lack of adequate police protection, further alienated the working middle class, and middle-class women hesitated to go downtown to shop. Each aspect of the problem fed upon the other parts.

The governments of American suburban communities quickly realized that they could live more economically if the houses were fairly large. Big houses contributed more tax money, and the occupant families generally demanded less public service than the small houses of the lower middle class. For one thing, large-house owners had fewer children and sent more of them to private schools. Since education was one of the most expensive items in a suburban budget, this could be quite a factor. In order to achieve an end desirable to them, suburbs erected zoning barriers against small and medium houses by requiring half acre, acre, or even larger lots. The lots were so expensive that it made no sense to put small houses on them, but the large lots took up more land than did compact development on smaller lots. Reciprocally, suburban "snob" zoning helped to concentrate the really poor in the central city, and this made the city's problems worse.

Gross disparities in tax burden and public service among the suburbs of major cities are characteristic of the United States—and almost unknown elsewhere in the world. Most countries fund education centrally, or at least the states do it, not the local towns. Police and fire protection are likewise centrally financed, as is most street maintenance. Only the United States reposes so much authority in local government and imposes on it such heavy financial burdens.

Canadian cities had a different history. Most expanded geographically to take in "outside" land as they developed, and few were "balkanized." Taxes therefore were averaged over large areas, and internal differences in tax rates and services did not induce internal movements.

INDUSTRIAL EXODUS

After World War II, industry began moving out of the cities. City locations had been preferred when raw material and finished products were hauled by team and rail, and when factory workers either walked to the plant or rode trolley cars. Most industry was then housed in multistory "mill-type" buildings, with brick-bearing walls and heavy-frame interior construction. Floors were rated around sixty pounds per square foot, and the allowable vibration factor was not high. Elevators created bottlenecks in the movement of goods about the plant. Power from a central source, steam or water, was moved about by systems of belts and pulleys, which were inherently inefficient, and also dangerous to arms and legs.

During and after the war, older plants modernized in a hurry, and the new processes were better adapted to sprawled one-story buildings on heavy concrete slabs which could carry almost unlimited weight and stand unlimited vibration. Belts and pullies gave way to individual electric motors. Small tractors moved materials about the plant without waiting for elevators and with considerable economy of manpower.

The new plants required big tracts of land, compared with the old ones which had been built from lot line to lot line. Assembling large in-town plots was frustrating and time consuming, and the plants found advantages in going out a distance and buying farms or portions of farms in one purchase. Goods now moved by large trailer trucks, and many modern plants needed good highway access more than they needed rail. Also many sites were available with both rail and highway access. The work force began to come and go by automobile, and this called for large parking lots. The plants also wanted elbowroom for further expansion and often bought a great excess of land over what was needed at the moment of purchase. Extra land was either landscaped to provide a pleasant view from the president's window, left to nature's own devices, or some of each.

The industrial exodus was almost complete by the mid-1970s. Only a few industrial plants remained in the cities, and these housed light industry which could function in the old mill buildings. The exodus had seriously reduced the urban tax base since the plants had been heavy tax contributors and low consumers of urban service. As the work force relocated in the suburbs near the plant, the combined move of plant and people exerted an important centrifugal force.

A curious by-product of the industrial movement was the isolation of the very poor in the central city. Most of the very poor were unqualified for much beyond manual work, and such manual jobs as survived the mechanization of practically everything were now located in the suburban plants, which the very poor, lacking automobiles, could not reach.

TRANSPORTATION SYSTEMS

For centuries cities depended on human feet. People walked where they were going, so distances were kept short and the cities were very compact. Cities relied on walls for defense, and this further restricted their size and increased their density. Until the nineteenth century, technology limited construction to a five- or six-story height, but within this limit a surprising number of people were shoehorned into a medieval city like Paris by economical land use.

The "age of feet" was followed by a brief but explosive period when the electric trolley car dominated urban transportation. The trolley erased distance, but it could go only where its tracks let it. Development followed the car lines with wide gaps at the urban fringe between the lines.

By the mid-twentieth century the trolley car had yielded to the automobile and bus. Like the human foot, the auto could go anywhere, and like the trolley, it could erase distance. The sprawled American city was as much the product of the automobile as of any factor. Without the automobile it could not have taken the sprawled form it has.

Human feet required streets and footpaths; the automobile required streets, and public bodies determined when and where to build them. But like the other forces that shaped cities, a "hen and egg" relationship existed between development and road building. Sometimes one led and the other followed; sometimes it was vice versa. Every major highway improvement affected the direction of development. For example, the George Washington Bridge across the Hudson, which opened in the late 1920s, overnight converted rural Bergen County, New Jersey, into a busy suburban area. The same thing happened almost every time a major bridge or tunnel gave access to an area hitherto hard to reach. Every major road permitting fast traffic to move to and from the urban core encouraged suburban development and aroused many areas from rural somnolence into suburban hustle. Penn Hills outside Pittsburgh, Pennsylvania, for example, was built up quickly after a major parkway linked it with the Golden Triangle.

The abundance of cheap gasoline was an ancillary factor. During the period of maximum urban expansion, gasoline cost about 25¢ a gallon, compared with $1.30 in early 1980. Air pollution in the 1940s concerned only industrial cities like Pittsburgh and St. Louis.

The automobile and bus were not yet considered prime offenders and became the mainstays of urban transportation. Buses remained a minor part of the system in early 1980 but became a growing factor as gasoline got more and more expensive.

Few fixed rail systems were built between the 1920s and the 1960s. The flexibility and convenience of the automobile reduced the extent to which existing railroads were used and discouraged new construction. But new systems began to appear in the late sixties and early seventies.

The Washington, D.C. Metro is an interesting example. Origin and destination studies suggested that it would not generate enough traffic to "pay for itself out of the fare box," but a group of stubborn planners, including the present writer, then chair of the National Capital Planning Commission, argued lustily in its favor, as did Mrs. James Rowe who followed as chair. The results surpassed expectations. The system has been heavily used, and new development has tended perceptively to drift toward transportation stops. In the late seventies a pattern began to emerge in many cities of short drives to outer fringe parking lots and the use of public systems from there to downtown. Some of the systems involved fixed rail like the Washington Metro; many more used express buses.

The Public Direction of the Forces That Shape Cities

THE MORTGAGE FACTOR

By the late seventies the mortgage market was in the doldrums, with interest rates too high for many purchasers. The cost of single houses went through the roof, and new construction lagged. New development was almost entirely limited to the outer suburbs where tracts, if expensive, were at least to be had.

The mid-seventies was a banner period for the erection of large condominium projects and conversion of rental units. Many new projects resembled the row houses built in Philadelphia and some other cities since colonial times. Most new condominiums of the seventies were erected in agreeable semirural surroundings and combined the convenience and economy of compact living with the attraction of green grass and flowers. They offered the buyer many of the financial advantages of ownership, but relieved him of the daily responsibility for shoveling snow in winter and mowing grass in summer. However, condominiums are not without problems, present and potential. The cooperative apartment houses of the 1920s had demonstrated this when the depression undermined them financially. However, the condominiums were the best game on the block in the 1970s, and sales were correspondingly brisk.

Rapidly rising costs, especially property taxes, and rent control, either real or threatened, induced the owners of many rental apartments to convert to condominiums; and many a once-happy tenant became a reluctant condominium buyer in order to avoid the need to move. This left a real shortage of rental space for those who either could not buy or had good reason to choose not to. Few new rental units were built in the late seventies.

Public influence could be increased very quickly if circumstances seemed to warrant it. Adjustment of loan underwriting rules could give preferences wherever policy dictated. The huge success of the "608 program" showed what *could* be done. A revised program of the Federal Housing Administration with help from an agency like "Ginny Mae" could reverse the financial vector to encourage building within urban areas, thus avoiding further outward expansion into farmland. These agencies could accept loans below the prohibitive commercial rates, for example, on new rental units. This force could also direct new development into foothills and other areas not adapted to farming but desirable for residences. Of all the factors readily subject to public control, the mortgage market is probably the easiest to influence without much new legislation or debate.

TAXES

Many ways could be found to equalize taxes between cities and suburbs, and among suburbs, but the political obstacles would be formidable. The states could assume such public functions as education, welfare, and police and fire protection, or the states could collect more revenue and redistribute it to local government on some equitable basis. However, these changes would require either legislation or constitutional amendment or both in most states and would meet strenuous local opposition. Local governments have always been jealous of their prerogatives, especially the management of their own schools, and they would yield this power only under considerable duress. States have hesitated to give funds to local governments "without strings."

The problem is peculiarly American; other countries have it only to a trifling extent. But this fact does not make the present property tax system any easier to modify. As mentioned previously, the property tax problem is a major element of difference between American and Canadian cities.

The income tax system could also be changed to extend to renters the same tax concessions that owners enjoy. This would require congressional action and would be opposed by powerful pressure groups. The tax factor, in other words, *could* be changed, but is not apt to be.

TRANSPORTATION

Urban public transportation in the United States has long been considered to belong in the private economic sector; and if not actually lodged there, it "ought to be." The rest of the world has felt otherwise. There, public transportation is regarded as a public responsibility, and buses, subways, and trains are run by the government. Most other countries have better public transportation than the United States; America's tolerance for third-rate facilities is amazing. Modern American buses are among the world's most sumptuous, but the supply and frequency of service leave much to be desired.

"Fare-box" mentality induces many transportation officials to restrict or eliminate runs which fail quickly to generate enough revenue to pay. This might occasionally be necessary, but a run should be left until it has proved that it will not "catch on." Sometimes in other countries consumers have taken from six months to a year to revise their riding habits.

By providing abundant, comfortable, safe, and convenient public transportation, government could considerably influence the course of future development. To do so would require accepting the fact that the fare box seldom has paid the costs of local transportation and that the property owners who benefit from it should help to pay for it through taxes. The tenants in tall buildings pay the whole cost of elevator service out of their rent, and the difference between horizontal transportation and vertical transportation is one of distance and not of basic concept.

The use of private cars, which the highway program encouraged through the years from World War II to 1980, could be discouraged in many ways. One would be a hefty tax on parking space in the city, making automobile commuting more expensive than buses or trains. The increased cost of gasoline in the late seventies had somewhat that effect in the free market. This, in turn, could be increased by withdrawal of all price control on gasoline, letting it seek its world market price. If this were not enough to discourage individual car use, the gasoline tax at the pump could be increased.

The city of Singapore recently took drastic action against the use of automobiles downtown. Special permits were required to drive an automobile inside the congested downtown zone, and the permits were expensive. The result has been that few cars enter this zone. Most people, car owners and others, use Singapore's excellent public transportation system.

Finally, public decisions about transportation could rechannel tax money away from roads and into better public transportation. Politically this would not be easy. Americans love their cars, and their attachment to them amounts to addiction. Highway lobbies are among the most powerful in Washington and the state capitals. Politics is the art of the possible, and it must be recognized that some of the ideas put forward in this section are politically unrealistic. However, each has been tried and each has worked somewhere in the world. This vector *can* be redirected.

Farming and Gardening in the Frazzled Urban Fringe

Green crop acres disappear for more than one reason. Some farmland has been abandoned because, given the 1980 state of agricultural technology, some crops cannot be grown in competition with the expansive fields of the Middle West. This is partly a function of size. Modern farm machinery has grown very large, and, like the giant dinosaurs, the large machines need room to maneuver. They cannot operate close to fences; small fields are not for them. Furthermore, they are enormously expensive; a small-scale operator cannot afford them. The little farms which once dominated eastern agriculture simply could not operate this way. They could, on the other hand, operate, and not inefficiently, with the scale of machinery that was in common use as recently as twenty-five years ago. Some manufacturers are beginning to make small models again, and the market for used equipment in good condition is brisk. Small machines are made in the Far East and increasingly used in Japan, Taiwan, and South Korea.

Expanding cities sent forth shock waves of rising land prices which discouraged many from continuing to farm. Older farmers took their profits and retired, rather than "holding on" and paying high suburban taxes. Younger people who wanted to farm either full or part-

time could not afford the high suburban prices. Much land, by-passed
by leapfrog development, was still farmed; some grew up in brush;
and some actually reverted to forest. By-passed land, temporarily
abandoned, could be returned to agriculture given adequate economic
incentives. Human inclination seems to be abundant; many eastern
young people want to farm, and small-scale farming can produce a
lot of food.

Land which is built on or paved over is another story. If used for
close-packed residential or commercial buildings or as parking lots
associated with either, all food-producing potential is permanently
lost. If, however, the land is developed as exurban plots of five or
so acres or as one-acre or half-acre subdivisions, then a real potential
remains for large-scale gardens; a surprising amount of food can be
grown in a well-manicured garden plot. When industry takes over
farmland, large spaces are often left around the factory and its parking lot and are either landscaped or "left raw." This land could be
gardened or small-scale farmed, with the loss of some aesthetic ame-
nity to the factory, but with the ability to grow a lot of food.

The Connecticut Agricultural Experiment Station reported in 1977
and 1978 on the productivity of gardens. The Day Waverly Gardens
in New Haven are on land reclaimed after buildings had been demolished and in 1976 produced about 450 pounds of vegetables per
1,000 square feet, worth $213 at 1976 supermarket prices. Experi-
mental plots in Lockwood Farm outside the city produced 1,120
pounds per 1,000 square feet, worth $425. Troy-Bilt, a manufacturer
of garden machinery, described in their 1980 advertising a plot of
1,200 square feet which produced in 1979 over 1,400 pounds of vegetables worth $912 by 1979 prices.

The United States Department of Agriculture reported annual per
capita vegetable consumption at 143 pounds, at which rate the Lockwood Farm test plots would keep about seven and a half people in
green groceries. If people ate twice as much in the way of vegetables
and half as much meat as in the late 1970s, a Lockwood-size plot would
go far toward feeding four people the year round.

In a moment of whimsy, some agricultural economists calculated
how much tomato juice the large backyards of greater Hartford could
grow if fervently cultivated. In 1980, towns with densities below the
county median covered 370,000 acres. Allowing for uneven terrain
and the reluctance of suburbanites to plow up their lawns, the potential garden space was arbitrarily estimated at one-fifth of the area of

these sparsely built towns. This would be about 74,000 acres or 3,225,000 garden plots of 1,000 square feet each. About 110 tomato plants could grow in each plot if set a yard apart and carefully staked. Under good conditions each plant would yield a bushel of tomatoes and produce about 15 quarts of tomato juice, or 1,650 quarts per garden. The result would have been an unthinkable 5.25 *billion* quarts of tomato juice, or 1.75 million cubic feet. An olympic-size swimming pool takes about 6,000 cubic feet, so this would have been enough to fill a great many pools. To pursue whimsy a step further, these figures suggest the virtues of developing automobiles that would run on tomato juice.

The example is silly, of course, but does suggest the productive capacity of America's suburban back yards. Back yard gardens are peculiarly American. Most countries are built too compactly to permit them. Australia and Canada have them, and England has small ones. Continental Europe has very few, and Asia, virtually none.

Until about 1960, farmland in the fringe was supposedly taxed on its market value. Assessors, mostly old-timers who knew the farmers personally, often went easy on the farms and did not adjust assessments upward nearly as fast as market conditions would have justified. Starting with Maryland in the 1960s, a movement spread throughout most states to tax fringe farmland not on market value but on usufruct. This idea was new to the United States, but ancient in Europe. It involved assessing farmland at what it would be worth as a farm, not as a potential urban site. This was the backbone of assessment systems in most of the German states before World War I. It did not entirely abandon the ad valorem concept but changed it to value for a specific use, not the "highest and best" (which means most profitable) use. This tax concession was a small step toward preserving land for farming, and at least it reduced the pressure on the farmer to quit farming and sell in order to avoid taxes which farm income could not meet.

Land Converted from Farming to Recreational and Conservation Use

Some of the land which went out of use for field crops went into recreational use as parkland or as golf courses. Rough terrain often went into park or conservation land, and since it was not good farmland anyway, its loss was not serious.

Some European countries like Holland and Denmark, which were short of land, had definitely articulated policies of preservation but diverted some land to either recreation or conservation or to both uses simultaneously. Since these countries are quite flat, the land they diverted was prime farm property. Recreational space is clearly an urban requirement; as cities grow larger, more such space is needed. Asian cities traditionally have lacked space, but even there some farmland has been converted to recreational use like golf.

What Can Be Done to Preserve Farmland

The last two chapters in this book discuss in detail what the uninhibited market could do and what regulatory steps have been taken in the United States and in other countries to respond to the threat of farmland loss. Several foreign countries have concluded that the free market forces will not do an effective job, and they have moved in with various programs.

Property tax relief, mentioned in an earlier section of this chapter, is only a partial response. Virtually all countries which seek actively to preserve farmland have gone beyond this. Most of them have zoned the rural areas, and conversion from farming to nonfarm uses requires permits from government boards. The United States is badly set up to undertake rural zoning, since this is regarded as a state, not a federal, power and since almost all the states delegate the power to local government. Many suburban towns are so dominated by developer interests that they are unlikely to zone any land away from development. Exurban towns, still dominated by the rural populations, contain enough people who are waiting greedily to unload their properties at great profit and enjoy their declining years in the Florida sun, so they are not inclined to pass anticonversion zoning laws either. If such laws are to be passed in the United States, it will have to be in state legislatures, and it will be fought by many local officials as an invasion of their prerogatives.

Some countries established heavy capital gains taxes on profits from the sale of land because the vendor himself had done little to "earn" the profit. Society engulfed his property; society, not the individual, made the market value rise; so society as a whole should profit rather than a single individual. The individual's participation in the process was often purely fortuitous; he happened to have made a prudent purchase long ago or he was lucky enough to have inherited a farm with urban potential.

Some American states and some foreign countries have experimented with public purchase either of the whole property, in which case it is leased to a farm operator, or else have bought the development rights and left the land to trade at or near its purely agricultural (usufruct) value.

Many states have taken steps which indirectly slow urban expansion and help preserve farms. State laws establishing subdivision controls, water and sewage requirements, and ecological impact studies tended to change the process of development from something an owner could do as a matter of right to something which was a privilege to be given or withheld by public officials. These state-imposed requirements clearly slowed the pace of development. On the other hand, when ventures that made as much money as successful estate developments became a matter of privilege which officials could give or deny, the specter of corruption intruded. Enough stench of corruption had been detected in the United States to induce the 1978 publication of a book on the subject by J. A. Gardiner and T. R. Lyons, *Decisions for Sale: Corruption and Reform in Land Use and Building Regulation.*

The future awaits the accumulation of enough experience by government trying to preserve farmland to determine the best course to follow. Meanwhile, the acute need is to formulate a national land policy. This is something that every land-short country has done along with several countries which once thought they had abundant land. Hopefully the United States will think through a prudent land policy for the 1980s and 1990s before a serious emergency forces it to contrive one overnight.

Brian J. L. Berry

2

The Urban Problem

Introduction

I was assigned "the urban problem" as the topic for this chapter, apparently in the belief that I would know what "the problem" is when discussed in the context of a book dealing with changes from agricultural to urban uses of the land. I must admit to a little skepticism about the consensus implied by the title. The thrust of my conclusions is that policies to curb farmland conversion may be more significant as tools of inner city revitalization than as methods of saving the land.

There clearly is growing popular concern about conversion of agricultural land into urban uses, and this concern is being translated into legislation in certain states. Yet even the Department of Agriculture is not without some ambivalence on the subject. To quote from a

BRIAN J. L. BERRY *is the Williams Professor of City and Regional Planning, director of the Laboratory for Computer Graphics and Spatial Analysis, and a faculty fellow of the Institute for International Development at Harvard University. Previously, he was the director of the University of Chicago's Center for Urban Studies and a member of the geography faculty. Dr. Berry is a member of the National Academy of Sciences and has been president of the Association of American Geographers. He has written more than 50 books and 200 professional articles.*

recent U.S.D.A. publication, *Major Uses of Land in the United States: 1974* by H. Thomas Frey:

> The growth of the urban area has traditionally been assessed in terms of its effect on agricultural land supplies. From this standpoint, urbanization is not a serious immediate problem except in some specialty crop areas. Probably not more than 35 to 40 percent of the 750,000 acres urbanized each year is cropland.

TABLE 1.   CHANGE IN SPECIAL-USE AREAS, UNITED STATES, 1969–74 *

Special-Use Area	1969 †	1974	Change
	Million acres		
Urban areas	31.0	34.8	3.8
Rural transportation areas	25.8	26.3	.5
Recreation and wildlife areas	81.4	87.5	6.1
Public installations and facilities	25.5	25.0	− .5
Farmsteads and farm roads	8.4	8.1	− .3
Total	172.1	181.7	9.6

* From H. Thomas Frey, *Major Uses of Land in the United States: 1974*. Washington, D.C.: U.S. Department of Agriculture, Agriculture Economic Report No. 440, November 1979.

† From H. Thomas Frey, *Major Uses of Land in the United States: Summary for 1969*. Washington, D.C.: U.S. Department of Agriculture, Agriculture Economic Report No. 247, 1973.

In light of this fact, my dilemma is this: am I talking about a genuine national problem of agricultural land conversion or a popular movement that has emerged from legitimate but highly localized concerns? To cast some light on the question, we must begin by coming to grips with demographic changes that have unfolded in the United States in recent years.

The Changing Nature of Urban Expansion

BACKGROUND DEMOGRAPHICS

Dominating all other national demographic trends is the continuation of a long-term decline in the rate of population increase. The population of the United States continues to grow, but at a steadily decreasing rate. During the 1950s, the national population grew 19 percent; during the 1960s, 13 percent; and if growth rates continued through the close of the 1970s, the nation's population in that time period increased 8 percent. Each of the components of population change—birth rates, death rates, and immigration rates—contributes to the current low degree of population increase.

The annual death rate, after falling continuously since 1900, stabilized during the 1950s at about 9.4 deaths per 1,000 and then dropped again to a 1979 level of 8.9 deaths per 1,000. The nation's birth rate has returned to its previous trend of long-term decline following the anomaly of the post-World War II baby boom. The birth rate stood at 19.4 births per 1,000 in 1940 and rose to 24.9 births per 1,000 in 1955—but has declined continuously since, dropping to 14.7 births per 1,000 by 1975, the lowest rate in American history. The eventual number of births that women now moving into their childbearing years are expected to have averages 2.1, a figure barely at the replacement level for a stable population.

Immigration rates are based on quotas that are fixed by law, and legal immigration currently averages 400,000 persons per year. Of the nation's total population increase of 1.7 million during 1975, 1.2 million resulted from natural increase (an excess of births over deaths), while immigration accounted for the remaining .5 million (including 130,000 Vietnamese refugees).

These declining rates of population growth have caused the Census Bureau to issue a new series of three population projections that adjusts "expected" national population growth downward. The projections suggest that the nation's population by the year 2000 may total between 245 and 287 million, significantly lower than projections made as recently as the latter half of the 1960s, which ranged from a low of 283 million to a high of 361 million.

The long-term decline in population growth is expected to continue, although previous fluctuations in birth rates will continue to affect current changes in the nation's demographic profile. The subpopulations of individuals aged eighteen to twenty-four and twenty-five to thirty-four, age groups now consisting of members of the postwar boom cohorts, have grown 13 and 23 percent respectively in the 1970s. The aging of these large cohorts, together with their greater life expectancies, will play a major role in increasing the median age of the nation's population.

Changes in the size of other age groups during the first half of the 1970s included a decline in the number of youths and an increasing number of elderly persons. These shifts in the demographic structure of the nation's population are occurring at the same time that fundamental changes in the overall structure of marital arrangements are emerging.

As the large birth cohorts of the late 1940s and early 1950s advance through young adulthood, the nation's marriage rate is now declining (after peaking in 1972); the median age at first marriage is increasing; the divorce rate is increasing (from 2.2 per 1,000 population in 1960 to 4.8 per 1,000 in 1975); more young unmarried adults are maintaining their own homes; and more children are living at home with a single parent. Since 1970, the largest increase in family groups has been among those headed by women who do not have husbands living with them; half of this increase was accounted for by women who are divorced. The combination of falling birth rates and changing household composition (especially the increase in one-person households) is reflected in the declining numbers of persons per household. Indeed, between 1970 and 1975, the number of primary family households increased 8 percent, and the number of primary individual households rose 30 percent. These different growth rates have reduced the proportion of households composed of related family members by 3 percent in just five years.

THE URBAN REVERSAL

Accompanying these demographic changes have been significant shifts in regional growth patterns. Signs of a shift away from the long-term trend of metropolitan growth exceeding that of nonmetropolitan areas first appeared in the 1960s. During this time, several nonmetropolitan regions experienced a reversal from population decline to modest increase, and it appeared that, in at least some of these areas, outward migration had peaked during the previous decade. The metropolitan population growth of twenty-two million during the 1960s resulted in part (one-third) from growth through the addition of new land area, but two-thirds was derived from population increases within the 1960 boundaries. Of this growth within earlier boundaries, three-quarters was due to natural increase; of the remaining quarter, a larger proportion resulted from immigration than from the inward migration of former nonmetropolitan area residents. Thus, only a small proportion of the increase in America's metropolitan population during the 1960s can be attributed to outward migration from nonmetropolitan areas.

Since 1970, a reversal has occurred; the growth rates for nonmetropolitan areas now exceed those of metropolitan areas. Nationwide statistics for the first half of the 1970s indicated that population had increased 6.3 percent in nonmetropolitan areas and only 3.6 percent in metropolitan areas (Table 2.).

TABLE 2. POPULATION OF THE UNITED STATES, 1950–1975 (IN THOUSANDS)

Residential Category	1975 *	1970	1960 †	1950	1970– 1975	1960– 1970	1950– 1960
					Percentage Change		
Total U.S.	208,683	199,819	179,971	151,235	4.4	13.3	19.0
Metropolitan	141,993	137,058	119,595	94,579	3.6	16.6	26.4
Central City	60,902	62,876	59,947	53,696	−3.1	6.5	11.6
Suburban	81,091	74,182	59,647	40,883	9.3	26.7	45.9
Nonmetropolitan	66,690	62,761	60,384	56,656	6.3	6.8	6.4

Sources: U.S. Bureau of the Census (1972), *Census of Population: 1970.* Washington, D.C., U.S. Department of Commerce. U.S. Bureau of the Census (1975), Mobility of the Population of the United States: March 1970 to March 1975. Series P-20. No. 285 in *Current Population Reports.* Washington, D.C.: U. S. Department of Commerce.

* 1975 data are April-centered averages from the Current Population Survey; 1970 data are also from the Current Population Survey and have been adjusted by excluding inmates of institutions and members of the Armed Forces residing in barracks for comparability with 1975 data.

† Data for 1960 and 1950 are total population counts from the two respective decennial censuses. The total population counts for 1970 were used to calculate the 1960–1970 percentage changes rather than the Current Population Survey figures shown.

When the nation's metropolitan areas are divided into their central city and suburban parts, it is readily apparent that the current lower rate of growth of metropolitan areas has resulted from a combination of the depopulation of the central cities and the slackening growth boom in the suburbs.

The central city population losses during the 1950s and 1960s were largely confined to the industrial heartland cities of the North Central and Northeast regions of the country. During the 1950s, 81 percent of the central cities that lost population were located in this northern area extending from the Midwest states through New England. During the 1960s, this concentration of declining central cities in the North lessened somewhat, to 74 percent.

In the 1970s, the greatest concentration of central cities that lost population continued to lie within this northern industrial area. But in the South, the central cities in metropolitan areas with more than one million residents also lost population, while the central cities of metropolitan areas with less than one million residents gained population—resulting in only a slight decrease in the total number of southern central city residents. In the West, on the other hand, the number of residents in metropolitan areas of all sizes increased, with the largest gain occurring in the central cities of metropolitan areas with less than one million residents (Table 3).

NONMETROPOLITAN GROWTH IN THE 1970s

Throughout the 1940s and 1950s, the nation's nonmetropolitan areas experienced high levels of outward migration. Some nonmetropolitan areas reached a turning point during the 1960s. In the 1970s nonmetropolitan areas as a whole not only retained residents but also experienced a gain in population through migration from metropolitan areas.

More significant for nonmetropolitan areas than their relatively higher growth rate was the *reversal* that occurred in migration between the nonmetropolitan and metropolitan areas of the nation.

Increased mechanization of farming since World War II had led to a decrease in the size of the farm population and contributed to rural outward migration. During the 1950s, nonmetropolitan areas experienced a net loss of more than five million persons through migration. High levels of outward migration continued into the 1960s, when the nation's farm population declined at an annual rate of 4.8 percent. In the 1970s, however, the farm population declined at an annual rate of only 1.8 percent, and the farm population was at an all-time low of 8.9 million, 4.1 percent of the nation's total population. With fewer outward migrants and increased numbers of inward migrants, nonmetropolitan areas had experienced net migration gains of approximately two million persons since 1970, thus reversing the trend of population loss that had existed since the 1940s.

These migration reversals occurred in almost all nonmetropolitan areas of the country. Generally, those areas located immediately adjacent to but outside metropolitan areas (which accounted for 52 percent of all nonmetropolitan residents) experienced the highest nonmetropolitan growth rates: a 4.7 percent increase from 1970

TABLE 3. PERCENTAGE CHANGES IN THE POPULATION OF THE UNITED STATES, BY CITY, SIZE, AND REGION, 1970–74

| Region | Total | All Metropolitan Areas | | | Metropolitan Areas | | | | Nonmetropolitan Areas* | | | |
| | | | | | >1,000,000 | | <1,000,000 | | | | 2,500– | |
	Total	Total	Central City	Suburb	Central City	Suburb	Central City	Suburb	Total	<2,500	25,000	>25,000
Total U.S.	4.1	3.6	-1.9	8.4	-3.8	6.4	0.3	11.5	5.0	5.0	5.7	3.3
Northeast	1.2	0.2	-4.7	4.0	-5.9	3.1	-1.6	5.5	5.1	-31.9	18.9	-9.7
North Central	1.3	1.0	-5.5	6.4	-7.1	5.5	-3.4	8.1	1.8	-1.0	4.5	-4.8
South	6.7	7.6	-0.1	15.7	-1.2	18.8	0.4	17.4	5.4	11.6	3.0	8.8
West	8.0	7.1	4.4	9.0	1.7	6.3	9.7	15.6	11.6	-2.6	4.5	32.7

Source: U. S. Bureau of the Census (1975). Social and Economic Characteristics of the Metropolitan and Nonmetropolitan Population: 1974 and 1970. Series P-23, No. 55 in Current Population Reports. Washington, D.C.: U.S. Department of Commerce.

* Nonmetropolitan areas in this table are groups of counties with either no place of 2,500 or more residents (<2,500), counties with a place of between 2,500 and 25,000 residents (2,500–25,000), or counties with a place of more than 25,000 but less than 50,000 residents (>25,000).

through 1973, compared with a 3.7 percent increase for counties not adjacent to metropolitan areas. In particular, nonmetropolitan areas whose residents are relatively more integrated into metropolitan labor markets experienced higher rates of percent growth (Table 4).

TABLE 4. GROWTH OF METROPOLITAN COUNTIES BY LEVEL OF COMMUTING TO METROPOLITAN AREAS, 1960–74

	Population			Percentage Change	
Level of Commuting *	1974	1970	1960	1970–1974	1960–1970
>19 percent commuters	4,372	4,009	3,655	9.1	9.7
10–19 percent commuters	9,912	9,349	8,705	6.0	7.4
3–9 percent commuters	14,261	13,497	12,805	5.7	5.4
<3 percent commuters	27,912	26,628	26,207	4.8	1.6

Source: Data supplied by Richard L. Forstall, Bureau of the Census.
* Percentage of counties' work force commuting to a metropolitan area for employment. Based on 1970 Census commuting data.

The areas of nonmetropolitan America that had undergone reversals from population decline to growth in the 1960s and 1970s were both diverse and widespread. In the South, an area extending from the Ozarks through eastern Texas that contains a predominantly white population had shifted from reliance on agricultural employment to development of manufacturing and new recreational areas. The Upper Great Lakes area, bordering the southern coast of Lake Superior, had also experienced growth primarily as the result of manufacturing decentralization and the development of recreational facilities and retirement communities. The nonmetropolitan areas of the Blue Ridge-Piedmont, Florida, the Southwest, and the northern Pacific Coast had all experienced growth resulting from the decentralization of manufacturing, recreational and retirement developments, the opening up of new resources, or the expansion of improved transportation facilities (for example, the interstate highway system) that enable persons to live in rural areas and commute to metropolitan labor markets. Indeed, the change was by 1980 so widespread that only one of the nation's rural areas continued to lose population through outward migration: the old Tobacco and Cotton Belt extending from the North Carolina Cape to the delta area of the Mississippi River. This area, which contains a large rural black population, had not benefited significantly from the decentralization of manufacturing and continued to lose residents through outward migration to cities of both the North and the South.

CONTINUING RESIDENTIAL DECENTRALIZATION

Amidst these major shifts, classical residential decentralization does continue, of course. As noted, growth rates were higher in counties with the greatest commuting flows into metropolitan areas. Moreover, between 1960 and 1970, the commuting areas of the nation's central cities themselves expanded. The expansion was most rapid in the southwestern Sun Belt; while a pattern of declining dependence of inner suburbs on central city job opportunities alongside continued, outer expansion of urban labor markets began to appear in the northeastern Snow Belt.

LABOR FORCE MIGRATION

Residential decentralization has been accompanied by similar shifts in the workplace of the labor force (Figures 1 and 2).

Comparing the national work force migration patterns of 1960–63 with those of 1970–73 reveals a dramatic reversal: once again, in the earlier period, the central counties of metropolitan areas gained 104,000 workers and nonmetropolitan counties lost 106,000 workers; in the more recent period, central counties lost 84,000 workers and nonmetropolitan counties gained 19,000 (Tables 5 and 6).

The Facts of Expansion Interpreted

What the foregoing statistics reveal is that since World War II there has been a breakdown of the nation's traditionally core-oriented settlement patterns on two scales. Interregionally, the heartland-hinterland organization of the economy as a whole is now giving way to a preeminence of the Sun Belt. Intraregionally, the center city is withering vis à vis the suburbs and the rural periphery, and indeed the withering of the core may be more of a problem than agricultural land loss at the edge.

This is, first of all, a result of the changing location of industry

Fig. 1. Net residential flows, 1970–1975 (in millions).

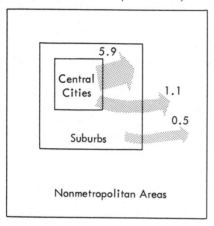

Note: Width of arrows is proportional to volume
of net flows among the three areas.

Source: U.S. Bureau of the Census (1975), Mobility of the Population of the United States: March 1970 to March 1975. Series P-20, No. 285 in *Current Population Reports.* Washington, D.C.: U. S. Department of Commerce.

Fig. 2. Net work force flows, 1960–1963 and 1970–1973 (in thousands).

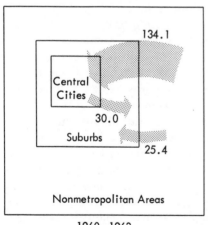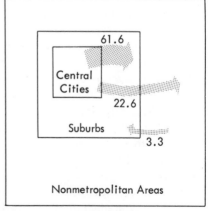

1960 – 1963   1970 – 1973

Note: Width of arrows is proportional to volume of net flows between the areas.

Source: Regional Economic Analysis Division (1976), Work Force Migration Patterns, 1960–1973. *Survey of Current Business.*

TABLE 5.   NET MIGRATION OF WORK FORCE FOR METROPOLITAN AND
NONMETROPOLITAN COUNTIES: 1960–1963 AND 1970–1973
(IN THOUSANDS)

	Metropolitan Counties Central Counties of SMSAs with Populations of:				Suburban Counties	Nonmetro-politan Counties
Years	2 million or more	1 million– 1,999,999	.5 million– 999,999	Less than .5 million		
1960–63	−27.1	71.9	28.0	31.3	55.4	−159.5
1970–73	−270.8	46.8	85.8	54.5	64.4	19.3

Source: Regional Economic Analysis Division (1976), Work Force Migration Patterns, 1960–73. *Survey of Current Business.*

Note: Positive numbers indicate net inward migration and negative numbers indicate net outward migration.

and of jobs. For the first half of the twentieth century, the northeastern Manufacturing Belt accounted for some 70 percent of the nation's industrial employment. Between 1950 and the mid-1960s, manufacturing jobs continued to grow in the Northeast, but the growth was more rapid in other regions of the country and the Manufacturing Belt's relative share fell to 56 percent. By 1970 relative decline had been replaced by absolute losses. From 1969 to 1977 the Manufacturing Belt lost 1.7 million industrial jobs, almost exactly the job growth of the former hinterlands. Similar shifts have taken place intraregionally. Between 1947 and 1958, to cite a few examples, central cities of the New York region lost 6 percent of their manufacturing jobs, whereas the suburbs gained 37.2 percent. In other heartland cities the figures are comparable.

Traditionally, the major central cities of the Manufacturing Belt were the centers of innovation. They were able to introduce new industries to offset losses of standardized industries to cheap labor areas elsewhere. But this is no longer true. The economy's rapid growth industries (electronics, aerospace, scientific instruments, etc.) are dispersed throughout the former interregional and intraregional peripheries; it is the older slow-growth industries that remain in the former cores. Employment in these remaining industries is extremely sensitive to cyclical change in the economy, which compounds the distress of northeastern central cities when the economy is in recession.

Even more critical is that the central cities of the former Manufac-

TABLE 6. PLACE-TO-PLACE NET MIGRATION OF WORK FORCE FOR METROPOLITAN AND NONMETROPOLITAN COUNTIES: 1960–63 AND 1970–73 (IN THOUSANDS)

	Metropolitan Counties								Suburban Counties		Nonmetropolitan Counties	
	Central Counties of SMSAs with Populations of:											
	2 million or more		1 million–1,999,999		.5 million–999,999		Less than .5 million					
	60–63	70–73	60–63	70–73	60–63	70–73	60–63	70–73	60–63	70–73	60–63	70–73
Central Counties of SMSAs with Populations of:												
2 million or more	—	—	0.1	73.7	—10.4	47.1	—3.1	20.7	65.1	75.8	—24.6	53.5
1 million–1,999,999	—0.1	—73.7	—	—	1.5	—3.2	—19.3	16.5	—23.7	12.6	—30.3	1.0
.5 million–999,999	10.4	—47.1	—1.5	3.2	—	—	0.6	—13.7	—1.2	—11.6	—36.3	—16.6
less than .5 million	3.1	—20.7	19.3	—16.5	—0.6	13.7	—	—	—10.2	—15.7	—42.9	—15.3
Suburban Counties	—65.1	—75.8	23.7	—12.6	1.2	11.6	10.2	15.7	—	—	—25.4	—3.3
Nonmetropolitan Counties	24.6	—53.5	30.3	—1.0	36.3	16.6	42.9	15.3	25.4	3.3	—	—

Source: **Regional Economic Analysis Division (1976), Work Force Migration Patterns, 1960–73.** *Survey of Current Business.*
Note: Positive numbers indicate net inward migration, and negative numbers indicate net outward migration.

47

turing Belt appear to have lost their traditional seedbed function.
The locus of innovation and growth has shifted elsewhere.

These job shifts have in turn precipitated the population shifts
already reviewed. Following the bulge in the population pyramid
formed by the post-World War II baby boom, there has been a decline
in fertility rates to less than replacement levels. As natural increase
diminished, migration has become a more important source of population change. This growing importance of migration as a factor of
growth has been intensified by the movement of the baby boom cohort
into its most mobile years. In all urban-industrial countries, a certain
minimum amount of geographical mobility is a structured part of the
life cycle. The greatest rates are associated with the stage at which
young adults leave the parental home and establish an independent
household shortly after formal schooling is completed. The baby
boom cohort is now passing through this stage, and the subsequent
period in which spatial differences in real wage rates and in employment opportunities provide signals that encourage economically motivated migration. This migration not only increases the well-being of
the movers themselves but also results in improved resource allocation. Thus, job shifts in a period of maximum potential mobility have
resulted in increased net migration from Manufacturing Belt to periphery for both majority and minority members of the U.S. population.

The South has experienced a dramatic and accelerating migration
reversal. Within regions, the balance of migration flows is away from
central cities to suburbs and exurbs and from metropolitan to nonmetropolitan areas. Throughout the nation, migrating workers have
left jobs located in major metropolitan cores for workplaces in smaller
urban areas, suburbs, and nonmetropolitan America. Since 1970, the
Northeast as a whole has lost population, a result of decreasing natural
increase and of the net migration reversal; in the South continued
high levels of growth have occurred, despite declining natural increase, because of increasing inward migration.

SHAPING FACTORS IN THE HOUSING MARKET

These outcomes have been shaped in important ways by the manner in which the nation's housing markets have operated and by the
influence of tax policy.

Urban growth has been strongly cyclical in both the long and short
term. American cities have not grown in a smooth or continuous man-

ner but in a series of major bursts, each of which has added a new ring of structures dominated by a particular building style. Nation-wide, the magnitude of housing investment probably has done more than any other factor to shape urban growth. The historical record of urban expansion in this century closely follows the peaks and troughs in the rate of capital formation in the housing sector. From 1910 to 1914 and again from 1921 to 1929, when real estate invest-ment boomed, metropolitan boundaries surged outward; later, when housing investment nearly came to a halt during the Great Depression and World War II, urban expansion slowed to a virtual standstill; then, in the 1950s, an unprecedented volume of housing investment was accompanied by a record pace of suburbanization.

Regionally, the expression of these cycles is to be seen in the differ-ent size of successive housing stock increments. Cities in older growth regions, especially those in the Northeast, have several growth rings with substantially differentiated housing stocks. Indeed, a good work-ing definition of the "inner city" is that area substantially constructed before the Great Depression. Cities located in newer growth regions have much more homogeneous post-World War II housing stocks.

Until World War II, less than half the nation's population owned their own homes, and less than half the housing stock was in single-family units—despite the fact that a national housing policy formu-lated in the 1930s sought to promote homeownership as a stabilizing social force. This policy relied heavily on new construction as a tool to upgrade housing standards and to provide for geographic mobility of urban households seeking better neighborhoods. The great surge of homeownership, however, did not come until the period between 1948 and 1960. It followed hard on the heels of the effective intro-duction of tax subsidies for owner occupancy, a by-product of the mass income tax adopted during the war.

In the United States, tax incentives are perhaps the chief instru-ment the government possesses for allocating investment resources among competing sectors of the economy. Housing investment reached its height between 1948 and 1960, partly because of the many new tax laws which singled out housing for favorable tax treatment at a time when household incomes were experiencing rapid real growth.

Historically, perhaps the most consistent bias in the federal tax code has been the favoritism given to investment in new structures relative to investment in the improvement and repayment of existing structures. This accelerates the rate at which buildings are replaced.

Although such speeding up of the replacement cycle for structures does not in itself give a locational bias to development, it compresses the period during which urban regions adjust to changed private market prices or new transportation technologies. It is to be noted, however, that when favoritism toward new construction is combined with other tax policies that favor homeownership, important locational effects result.

It is a peculiarity of the tax subsidy method of investment stimulation that the value of the tax advantage granted to homeownership is proportional to the marginal tax bracket for investors who claim the homeownership deduction. This deduction took an allocative significance for the first time during World War II when the marginal federal tax paid by most Americans rose from 4 to 25 percent, making the deductibility of homeowner expenses far more valuable than they previously had been. At upper-middle-income levels, the tax system creates an annual cost advantage for owner occupancy of about 14 to 15 percent. Little wonder, then, that the percentage of families owning their homes jumped from scarcely 40 percent in 1940 to over 65 percent in 1960.

This growth cannot be attributed solely to federal tax policy, of course. A number of other national policies—such as the introduction of FHA financing in the early thirties, VA financing after World War II, and the opening up of the suburbs through highway construction in the fifties—lent force to the homeownership boom. Household income growth also contributed in major ways to the higher rates of owner occupancy. Together, both directly and indirectly, these forces encouraged low-density single-family living patterns, with generous amounts of land consumption on the urban fringe. On average over the period 1950–70, each newly constructed single-family home added approximately six-tenths of an acre to the nation's urbanized area.

I find it difficult to believe that this rural land consumption is in any way "the urban problem." Rather, what has been facilitated is a restructuring of the nation's settlement patterns, and what has been left behind is the more significant problem.

THE COUNTER-URBANIZATION PROCESS

We now are witnessing the emergence of a new scale of low-slung, far-flung metropolitan regions and a new force of counter-urbanization: the transfer of the locus of new growth to some of the most remote and least urbanized parts of the country. This trend has been facilitated by job shifts and by a successful national housing policy which is oriented to promoting household wealth through homeownership, through improved living conditions via new construction, and through increased efficiency by means of mobility.

I have noted that the settings where this growth is now occurring are exceedingly diverse. They include regions oriented to recreation in northern New England, the Rocky Mountains, and the Upper Great Lakes; energy supply areas in the northern Great Plains and southern Appalachian coal fields; retirement communities in the Ozark-Ouachita uplands; small manufacturing towns throughout much of the South; and nonmetropolitan cities in every region whose economic fortunes are intertwined with state government or higher education. Among other factors contributing to these shifts are changes in transportation and communications. These have removed many of the problems of access which previously served to constrain the growth prospects of the periphery. They permit decentralization of manufacturing on the inexpensive land and benefit from the low wage rates of nonmetropolitan areas. Also contributing is the trend toward earlier retirement. This has lengthened the interval during later life when a person is no longer tied to a specific place by a job. Finally, there is increased orientation at all ages toward leisure activities caused in part by rising per capita income and centered on amenity-rich areas outside the daily range of metropolitan commuting.

These are but symptoms of the more profound forces which are at work, however. The concentrated industrial metropolis developed because proximity meant lower transportation and communication costs for those interdependent specialists who had to interact with each other frequently or intensively. One of the most important forces contributing to counter-urbanization is the erosion of centrality by time-space convergence. Virtually all technological developments of industrial times have had the effect of reducing the constraints of geographical space. Developments in transportation and communications have made it possible for each generation to live farther from activity centers, for these activity centers to disperse, and for information users to rely upon sources that are spatially more distant yet temporally more immediate.

In other words, large dense urban concentrations are no longer necessary for the classical urbanization economies to be present. Contemporary developments in communications are supplying better channels for transmitting information and are improving the capacities of partners in social intercourse to transact their business at great distances and at great speed.

The time-eliminating properties of long distance communication and the space-spanning capacities of the new communication technologies are combining to concoct a solvent which has dissolved the agglomeration advantages of the industrial metropolis, creating what some now refer to as an urban civilization without cities. The edge of many of the nation's metropolitan systems has now pushed 100 miles and more from declining central cities. Today's urban systems appear to be multinodal, multiconnected social systems sharing in national growth and offering a variety of life styles in a variety of environments. And what are being abandoned are those environments that were key in the traditional metropolis-driven growth process: the high-density, congested, face-to-face center city settings which are now perceived as aging, polluted, and crime-ridden with declining services, declining employment bases, and escalating taxes. These latter environments are the real urban problem.

At this juncture, we must face the question as to whether rising energy costs are likely to turn the equation around and precipitate recentralization. On this issue, I am one of the doubters, for several reasons.

1. The first-order effects are already being seen in the rapid changes in the United States auto fleet—changes which are likely to enable us to achieve the Environmental Protection Agency's gasoline consumption targets for 1983–84 one or two years early. The changes have already brought Chrysler to its knees and have enabled many Americans to more than halve their gasoline consumption for a given mileage.

2. Recentralization demands that workplaces relocate to the center once again. I have no doubt there will be reshuffling of residences closer to work over the years, but so long as changes in communications and production technologies permit dispersed workplace locations, and so long as avoidance of the negative externalities of the core remains profound, continued growth of smaller outlying towns and cities seems to be more likely.

3. The overall high rate of inflation, combined both with recession and the aging of the baby boom generation, will result in much lower mobility in the future and, therefore, less willingness and capability to readjust rapidly. Such major readjustments as continue to take

place will, I believe, be in the direction of environments which significantly reduce home heating costs, given that the first-order reduction in gasoline consumption can be made by changing the vehicle one drives.

In short, I do not believe that the current energy problems should be looked at as a possible corrective for outer growth and inner decline.

OUTER GROWTH AND INNER DECLINE: THE LINKS

Outer edge growth and inner neighborhood decline are intimately related through the straightforward link that exists between new housing construction and inner city abandonment. Since the early 1960s, new housing construction has far exceeded household growth. Between 1963 and 1976 household expansion was some seventeen million, but twenty-seven million new housing units were constructed. The crucial role of housing construction in excess of household growth in determining the value and maintenance of older housing is quite obvious. When new housing is built and occupied, more often than not by relatively well-to-do families, the older housing vacated by these families provides homes for lower-income families. If there is excess construction, the least desirable housing will be left vacant after this game of musical chairs comes to an end. Ultimately these buildings will be abandoned and demolished, often being the oldest and the most outmoded remnant of the earliest housing cycles.

Thus, periods of abundant housing investment are years of vigorous new housing construction, and in the United States most new housing always has been built near the urban periphery. But there is a natural tendency of a highly fragmented and speculative building industry to overshoot. Since 1960 more than one-third of all new housing construction has replaced older stock rather than added to the total supply. As a consequence, the high rate of housing production has not only pushed the urban boundary outward but has also produced faster abandonment of older housing, especially that located in the inner cities of the Manufacturing Belt metropolitan regions.

Seen in this light, abandonment may be viewed as a perverse measure of the success we have achieved in national housing policy and in preserving and enhancing the mobility so essential for the realization of new growth opportunities. Seen from another viewpoint, it is a by-product of overbuilding and excessive land consumption at the outer edge. Perhaps for this and other reasons, the inner city cannot and should not be so lightly written off. There is a clear national interest in the problems of the central city, derived from three distinguishable premises.

1. It is socially wasteful to underutilize, abandon, or destroy capital investments made by preceding generations in urban infrastructure, housing, places of business, and public buildings. The national product is diminished by our present course of reproducing these facilities elsewhere rather than using what already exists.

2. Although their populations are decreasing, central cities still hold a very large number of people whose lives and fortunes are unfavorably affected by their physical and social environments. Even assuming that suburbanities and exurbanites entirely escape these adversities, more than a fourth of the nation's people live in central cities and contend daily with their stresses. The welfare of these people is surely a matter of national interest.

3. Although central cities, as large, dense concentrations of people and jobs, have become technologically obsolete, the shift to a new spatial organization can perhaps be made less painful to those who must adapt to it. Orderly change would be less costly to society as a whole than allowing stresses to accumulate within a system trying desperately to maintain itself until the system as a whole fails.

National concern about these problems has been evident in a series of federal programs designed to reverse or at least to slow down the decay. Over a period of time, these programs have reflected a widening perception of the interplay of forces that cause central city deterioration. They have tested a variety of remedies, and they have spent many billions of dollars. All this, to date, has been of little avail, largely because funds have been poured into the problem areas rather than being applied to the problem of overbuilding at the periphery. Where such overbuilding is most constrained, inner city revitalization has been most likely to occur.

Private Market Inner City Revitalization as a Countervailing Force

In some neighborhoods in certain central cities, especially where suburban growth management has taken hold and as rising energy costs and recession have joined to halt real income growth, private market renovation has begun to produce a modest inner city resurgence, apparently running counter to the dominant forces of decentralization and dispersion. During the sixties, privately renovated neighborhoods such as Georgetown in Washington, D.C., Greenwich Village in New York, and Boston's South End were considered to be unique. But more recently, homeowners have begun to renovate old neighborhoods all across the country.

Part of the reason for revitalization is certainly the downturn in new housing starts since 1974. The metropolitan areas with the lowest rates of replacement supply have been those which, in general, have experienced the most rapid inflation of housing prices and the most substantial neighborhood revitalization: those with the greatest replacement supply have had the least revitalization.

But there is another source of variation among metropolitan areas, on the demand side. Since 1970 there have been many changes affecting the number and nature of new households entering the housing market. An increase in owner-occupancy rates from 60 to 75 percent between 1970 and 1975 has been accompanied by rapid change in the nature of baby boom generation households.

The increase in homeownership, occurring most rapidly among one-person or single-parent households, took place simultaneously with the massive inflation of housing costs. It surely represents an investment rather than a consumption decision: an inflation hedge against being priced out of the market in the future. The inflation, in turn, has been both cause and effect of the slower pace of new housing starts which of itself would have made inner city reinvestment more attractive.

Overlaid upon this have been the housing preferences of new higher-income young homeowners not pressed by child rearing. Often two workers, one or both of whom may be a professional, seek neighborhoods in the inner city with geographic clusters of housing structures capable of yielding high-quality services; a variety of public amenities within safe walking distance of these areas, such as a scenic waterfront, parks, museums or art galleries, universities, distinguished architecture, and historic landmarks of neighborhoods; and range of high-quality retail facilities and services, including restaurants, theaters, and entertainment.

These preferences follow directly from life style and compositional shifts. Continued development of American society has resulted in increased economic parity for women. This enables them to have the option of roles other than that of housewife and mother. In consequence, men and women lead more independent lives and are able to exercise more options in life course transitions. Increasing numbers of couples live together without the formal ties of marriage. The cost of child rearing is rising. Birth control technology has improved, and abortion laws have been liberalized. Hence the birth rate is dropping. The result is an increasing number of families with two or more workers. Working wives are more numerous than ever before.

Revitalization, then, has been taking hold first in superior neighborhoods in some of those metropolitan areas which have the lowest rates of replacement supply (i.e., have experienced *least* overbuilding and therefore relatively little dispersion). That is especially true in areas which have a sizable cluster of professional jobs that support the youthful college educated labor force most likely to evidence life style shifts. It is to be noted, however, that significant revitalization may be limited to the metropolitan centers with agglomerations of postindustrial management, control, and information processing activities, especially to those with rapidly growing downtown office complexes.

There is a growing preference for apartments, row or town houses, and innovative types of design, along with experimentation in forms of tenure, such as condominiums and cooperatives. These preserve some of the tax advantages of ownership and yet provide greater liquidity. Also popular are new forms of contracting arrangements for the operation and maintenance of housing. Such developments have occurred because smaller households require less space, because the fluidity of households and the looser legal links among their members are contrary to the rigidity of tenure associated with ownership, and because the maintenance of house and grounds is time-consuming. With most residents going out to work, domestic chores must be reduced to a minimum.

Central locations in the core city and the older suburbs have become increasingly attractive. There one finds an appropriate stock of housing, along with access to services and convenience in getting to work. And since many of today's households have no children, the racial factors of school integration do not act as they once did in producing the white flight to the suburbs.

But there are polarizing effects here as well. Over the past twenty years, the housing of the welfare poor and the working poor has improved primarily because they have fallen heir to what used to be called "the gray areas." The softening of middle-class demand for this housing stock lowered its relative price and permitted a sharp decline in overcrowding for low-income people. Whatever the troubles of the cities, this has been a fortunate outcome. But the danger appears imminent that the housing stock available to the working and welfare poor will now be sharply diminished, squeezed between reduced rates of filtering at one end and the childless multiworker household at the other. There is, in this, an incipient class conflict between the new young, well-educated, professional class, actively pursuing alternative living arrangements and life styles, and the majority of the children of working-class Americans for whom marriage and the home in the suburbs remain a desirable goal: Harris polls continue to report that of the 35 percent of American dwellers who plan to move in a two- to three-year period, 53 percent of them plan to move to a suburb or to a rural area. Indeed, one may argue that the main value struggle in the United States today is between liberal-regarding upper-middle-class intelligentsia (a minority of whom are our most vocal Marxists) and a newly middle-class and more conservative proletariat which rejects both "alternative" life styles and the left wing's egalitarian arguments. For the intelligentsia, material goals may have been succeeded by those of the quality-of-life and of self-actualization (for the Marxists, notoriety and in-group admiration); for the middle-class workers, however, material welfare and economic progress remain the dominant concerns.

Is it possible, then, for private market revitalization to be sustained? A variety of factors suggests otherwise, barring a powerful new approach to growth management and land conversion. First, the baby boom generation will age and be followed by much smaller age cohorts. The population bulge, as it ages, will continue to disrupt one national institution after another. In the 1950s and 1960s it created problems of expansion for the public schools and universities—institutions which more recently have had to cope with the ordeal of shrinkage as their user populations have subsided. When the population crest reached the eighteen to twenty-four bracket, it multiplied crime rates and redirected national job creation efforts to the alleviation of youth unemployment.

Perhaps the greatest adjustments for public policy of all kinds lie ahead, when the babies of 1950 become the aged of the year 2015. Among those adjustments, as before, will be those in housing preferences, causing the demand for inner city living in revitalized neighborhoods to subside rather than increase, at least until the early twenty-first century. Additions to the supporting job base may also be lacking. There appears little likelihood that new postindustrial employment clusters will develop in high-rise office complexes in inner city locations in the near future as they did in the 1960s; headquarters decentralization now appears the stronger force. Further, as the cohort ages, mobility will decrease. For all of these reasons, the prospect for wider inner city revitalization appears to be bleak unless continuing recession cuts deeply into the new housing industry or unless it becomes increasingly difficult to convert new land to urban uses at the periphery. If that comes to pass there will still be an inner city problem of the displaced and the disadvantaged.

The continuing public policy problem will then be that of ameliorating the heavy social costs incurred by the concentrations of captive individuals without access to the real economy: concentrations which continue to be characterized by high unemployment rates, especially among minorities, and by low educational achievement, drug addiction, crime, and a sense of hopelessness and alienation from society. Attention to causes rather than symptoms demands that factors which exclude individuals from the mainstream of society and from meaningful work opportunities be of prime concern, for work is a measure of worth in the U.S. Such factors include poor skills, cultural gaps, language barriers, and race. These may have to be addressed by both law and remediation. Programs aimed at creating work situations, intensifying quality educational assistance, and improving health and nutrition so that children will be able to learn are essential. Many such efforts at human capital development must be people-directed, but a deliberate strategy of fostering geographic as well as social mobility should not be excluded.

After World War II, a restructuring of incentives played an important role in the increase in homeownership and the attendant transformation of urban form. There is no reason to believe that another restructuring could not be designed to lead in other directions. In a highly mobile market system nothing is as effective in producing change as a shift in relative prices.

There is, then, a way. Whether there is a will is another matter. Under conditions of democratic pluralism, interest group politics prevail, and the normal state of such politics is business as usual.

The bold changes that followed the Great Depression and World War II were responses to major crises, for it is only in a crisis atmosphere that enlightened leadership can prevail over the normal business of politics in which there is an unerring aim for the lowest common denominator. Nothing less than an equivalent crisis will, I suggest, enable the necessary substantial inner city revitalization to take place. Until that crisis occurs (and I leave open the question of whether OPEC has engineered such a crisis), dispersion and differentiation will prevail. Some limited private market revitalization will continue, to be sure, but within a widening environment of disinvestment manifested geographically in the abandonment of the housing stock put into place by earlier building cycles.

Robert C. Weaver

3

Coordinating Rural and Land Policy

My orientation is one of concern for urban America and the continuing absence of a coordinated rural and urban land policy. While recognizing the potential dangers of too rapid and indiscriminate conversion of farmland to urban use, I leave speculation about the immediacy and dimensions of that problem to others. My emphasis will be, among other things, upon the impact on urban patterns of living and housing resulting from conversion of farm and other rural land.

For decades the nature and tempo of such land conversion has significantly affected urban forms, quality of life, availability of residential building sites, and the amount, location, and type of urban housing. Not only do land requirements for shelter, transportation, infrastructure, and industrial and commercial facilities set upper and lower limits on the acreage involved in conversion, but the efficacy of the resulting urban uses determines the actual amount of land that is

---

ROBERT C. WEAVER *is a distinguished professor emeritus of Urban Affairs at Hunter College, The City University of New York. He has been in public service for forty-seven years, including administrator of the Housing and Home Finance Agency under President John F. Kennedy and secretary of the Department of Housing and Urban Development under President Lyndon B. Johnson. Dr. Weaver has written four books and over 175 articles. He holds thirty honorary degrees.*

converted. And as Brian Berry points out, the availability of land for urban use and the amount of mortgage financing in suburbia and exurbia have a crucial impact upon cities.

Effective action to reduce sprawl and scatteration, avoid excessive minimum land requirements of single-family housing, and relax impediments to multiunit residential construction would produce several results. It would diminish the rate of farmland conversion, and it would promote more efficient and equitable patterns of urban land use. As we face a future of declining rates of population growth, possible reduced growth of suburbia, probable continuing population decline in cities, and rising rates of growth in exurbia, there are speculations of dramatic changes in urban America. The upgrading and rehabilitation of housing in some neighborhoods of some central cities is seen as the harbinger of these and other cities' revitalization. Others see this as evidence of the displacement of minorities and the poor by the more affluent.

In order to evaluate the impact of recent developments in urban land use, it is necessary also to consider what is transpiring and attempt to outline possible future trends. Perhaps the most crucial issue is whether or not the forces impinging upon urban America will continue to encourage sprawl or nurture more compact suburban clusters as well as more viable central cities. Related to this is the probable continuation of recent socioeconomic patterns and their impact upon urban America.

If exurbia should continue to expand, as the suburban growth rate declines, what will this mean for the cities? What of restrictive zoning? What of geographic racial residential patterns? And how will all of this affect efforts to house the poor and disadvantaged?

Certain current changes, such as the shortage and high cost of energy, are difficult to translate into institutional trends. Some observers do not believe high energy costs will cause urban recentralization. They cite the design of automobiles which will consume significantly less gasoline and our failure so far to relocate workplaces in the central city. Others assert that industries do not consider transportation or heating and cooling costs as critical factors in planning location. They add that households can insulate their homes and change transportation habits without moving into high density areas. Still others see continued concentration of jobs in nonmetropolitan locations as enabling people to live there and still easily commute to work. Yet there is much opinion that if and when the energy crisis abates, former patterns of urban development will be reestablished.

The high cost of housing in suburbia and the shift of job expansion beyond metropolitan areas will probably lead to population growth in exurbia. Land use patterns there will probably be like those which characterized suburbia during its period of great expansion. Considering the remoteness, size, and capacity of local governments involved, sprawl and scatteration may be even greater there than in suburbia.

Although the national rate of population growth is declining, the number of households is growing appreciably and new living and socioeconomic patterns are gaining prevalence. More households are composed of unmarried couples living together, childless couples, and one person. Even more significant is the rapid entrance of more women in the labor market and a spectacular growth in the number and proportion of married couples in which both husband and wife work. These new life styles have already expanded and modified the demand for housing and increasingly it is being satisfied by rehabilitation in the central city. Despite assertions to the contrary, there has *not* been a widespread return of middle-class Americans to core areas; rather, their exodus from cities has slowed down.

Some attribute this to the energy crisis. Others believe that it is due to inflation, the drastic rise in housing costs in suburbia, and soaring mortgage interest rates. They expect significant suburban expansion once the homebuilding industry revives and believe that many households which are now originating in the city will leave for suburbia. Others believe that the new socioeconomic living patterns will occasion a continuing growth of middle-class household formation in the core areas. I tend to lean in this direction, while recognizing that such growth will be limited to localities having a supporting job base.

The current neighborhood revitalization raises a social problem. It involves displacement of lower-income, often minority, households by more affluent homeowners or renters, the majority of whom are white. This is part of the larger urban problem of providing housing for lower-income people.

Many complexities and some paradoxes are involved. In the central cities, where the less affluent have been and remain increasingly concentrated, supply-oriented subsidized residential construction and deep rehabilitation expanded in the 1960s and upgraded the quality of shelter available to lower-income groups. It also reversed the penchant for urban renewal to displace the less affluent and the black and at the same time provided greater opportunities for these groups to find housing in some suburban areas. As middle-class and an increasing number of white middle-class families move out of the central cities, better housing becomes available to the residents remaining in core areas. Conversely, as more affluent households remain in the central cities, their presence slows down the upgrading of housing for the disadvantaged.

In this context there is much controversy over where housing subsidies should be concentrated and whether they should be demand-oriented or supply-oriented, and unfortunately, the discussion tends toward categorical positions. The true issue is not whether to provide rental housing for disadvantaged residents in the central city or to open the suburbs to them; nor is the issue supply-oriented or demand-oriented subsidies. Both approaches are needed and the mix will vary over time and from one urban area to another and, indeed, often from one neighborhood to another.

When in 1948 I wrote a detailed analysis of racial segregation and housing, my main emphasis was upon equal opportunity, free access, and removal of color differentials in the buying power of a dollar spent for shelter. Since then, the marginal federal income tax paid by most Americans increased over five times, and the tax advantages of homeownership became real for many more people. Not owning a home often meant losing out on economic benefits like favorable income tax treatment; a hedge against inflation; an often involuntary, or at least unconscious, form of savings; do-it-yourself maintenance; and low interest rates because of federal mortgage insurance and special advantages provided to mortgage-lending thrift institutions. Discrimination which limited blacks' access to homeownership by barring them from the suburbs extracted a real economic toll.

As industry and, more recently, other sources of employment shifted from the central cities to suburbia, racial exclusion lessened job opportunity for blacks. Not only were the locations of growing employment costly to reach in terms of time and money, but often remoteness was tantamount to inadequate knowledge of available employment. When firms moved to the suburbs, black employees often found it difficult to follow them because they could not find shelter in the new location.

Even if the "white noose" around the suburbs were penetrated, it would be a long time before many in the ghetto would escape. On the other hand, experience to date suggests that moderate-income black families who have skills or work experience and those with higher income are the ones most likely to seek suburban housing to enhance their job opportunities. They are joined by other middle-class blacks who work in the city and seek to upgrade their housing.

Between 1970 and 1977 the black population in Washington suburbs doubled, leaving only 57 percent of the blacks of the metropolitan area in the city of Washington; 76 percent had been concentrated there in 1970. The reasons for this exodus were indistinguishable for blacks and whites. In 1977 the household incomes of suburban blacks were, on the average, some 40 percent higher than those in the center city, and only slightly lower than the average incomes of suburban whites. The impact of the loss of middle-class blacks upon the economic base of the city was offset by the formation of new households—young professionals, single and alone, or single and living together, of whom a third were black and most middle-income. In an earlier period, black suburbanites had increased at twice the general growth rate in five additional cities: Atlanta, Boston, Chicago, Los Angeles, and Philadelphia. Movement of blacks to suburbia is, and will continue to be for some time, predominantly a middle-class phenomenon.

Many black residents in the central city are apprehensive about the treatment they would receive in most of suburbia. Those with few skills are realistic enough to assume that, by choice or lack of opportunity, they will remain in the central city for some time. Thus they opt for its rehabilitation and upgrading. This does not dictate, as some would advocate, a central city strategy which would ignore the opening of the suburbs and the provision of lower-income housing therein.

From the black perspective, a more sophisticated approach is one that recognizes the intrinsic locational potential of many black ghettos, urging that blacks apply entrepreneurial talents to control, occupy, and develop them. This position does not necessarily imply that all blacks are to remain in the central city. Many recognize the economic and psychological costs of exclusion from the suburbs where new housing has been concentrated and new jobs are still concentrated. They would agree with Congressman Parren J. Mitchell when he said, "Let's get the record straight. I'm not sure that most blacks want integration. I am sure that we want equal access to the housing market."

One thing is certain: if "regentrification" continues, that is if the upper middle class continues to move back to the city, there will be a need for either relocation or greater housing subsidies to enable the indigenous population to stay where it is. As in urban renewal, this suggests the need for larger programs of low-income housing on- and off-site, as well as demand- and supply-oriented subsidies for low- and moderate-income housing in affected areas.

If "regentrification" diminishes as residential construction revives in the suburbs and expands in the exurbs, the city population will be further eroded. This, in turn, will restore a degree of looseness in the housing market and demand-oriented subsidy programs can translate this into further upgrading of shelter for the less affluent. Demand-oriented housing subsidies, however, have not recently widened neighborhood choices for the poor in general and for blacks in particular, nor has the rate of movement to the suburbs been greater for recipients than for those who were not. Demand subsidies reflected the black-white difference that existed before they were introduced.

If nonmetropolitan population growth should continue to exceed metropolitan rates, many older suburbs will be adversely affected, and some older suburbs will look to new user groups to survive. Included will be those who work in the suburbs but cannot afford housing elsewhere. There will be economic impetus for relaxed prohibitions against the construction of multifamily housing and this will be reinforced by court decisions mandating easier zoning restrictions. Statutory requirements and HUD regulations tie Community Block Development Grants to provision of lower-income housing. As time passes, many older suburbs will have to comply with these requirements in order to get increasingly desired and needed federal financing for community facilities, but this will be a fitful process, for the commitment to exclusion in most suburbs is deep.

As progress was made in upgrading the quality of housing for low-income urban households, a long existing social problem became increasingly apparent. It was the perplexing problem of housing for the most troubled elements in the population. Few, if any, federal direct housing subsidy programs have faced up to it.

One proposal envisioned the emergence of a new breed of small-scale indigenous owners in the ghetto. They were expected to select tenants carefully as public housing did in its early years. Successful sponsors of Section 221(d)3 and Section 236 rental developments, in an understandable effort to attract the "respectable poor," tried to select carefully. On those relatively few occasions when public housing was able to break out of ghetto locations, management assured reluctant communities that carefully chosen families and the elderly would be selected. But what of families and individuals who presented serious problems to neighbors and management alike? Were they to be concentrated in public housing where they could be inundated by social services and special assistance, or were they to be widely dispersed in order that they would not feed on each other's problems?

Occupants who presented serious management problems, or were thought to do so, were the undesirables in the housing market. In New York City private landlords who accommodated such tenants have been found to charge exorbitant rents or provide minimal services or both. Events in the last decade have clearly demonstrated that housing alone cannot solve all the problems of all of those who are poverty stricken, alienated, and often least able to participate effectively in the urban labor market.

If Lee Rainwater was correct in his 1970 study of the Pruit-Igoe project in St. Louis entitled *Behind Ghetto Walls*, the most important need of poor families is more money. They certainly need more and better goods and services, and in a relative sense, their poverty needs to be made less striking. Housing subsidies are ill-equipped to meet the total economic need.

A return on the part of public housing to careful selection of middle-class-oriented poor may make the total program more viable, but it will only complicate the status of the rejects. We must either develop effective programs for the deprived in our society or adapt public housing to include them, recognizing that resulting shelter is only a small element in the overall problem. Either route is expensive. As Professor Eugene Meehan demonstrates in his recent study of public housing, many of that program's present difficulties are due to inadequate financing which led to "little concern with the social, psychic, or even economic dimensions of the tenants' lives." The late 1960s demonstrated that we cannot ignore the human beings involved and should cease the pretense that our housing programs do, or will, meet most, if not all, of their needs.

It is misleading to identify low-income or welfare households with "undesirable tenants." Many very poor households are model tenants while some quite wealthy people can be unbelievably obnoxious. A study of management of federally assisted housing in New York City, while recognizing the impact of tenant selection upon occupants, management, and the community, found "no necessary relationship between income levels and the incidence of rent delinquencies, vandalism, crime, tenant-management conflict, or turnover." It did identify drug addiction among tenants or in the neighborhood as the most serious social factor leading to maintenance and management problems. And it found that tenants of these projects often complained about the presence of addicts. Families with many children also caused maintenance and management problems; the latter could largely be avoided if the development were properly designed and the management adequately skilled.

There is encouraging evidence that we are becoming more sophisticated about this baffling problem, recognizing, for example, that a mere handful of young, uncontrolled tenants can create havoc in a large, multistory and multiunit development. Increasingly, we are questioning the feasibility and desirability of vast high rise projects for low-income families. Federal policy now discourages such architecture and stresses smaller developments for occupancy by households with children. Equally important, we have learned the relevance of design of low-income housing, property maintenance, and security.

Often where socially-motivated sponsors and managers have attempted to handle a few greatly troubled families, they have found the problem more difficult than they anticipated. In the words of Duncan Elder, the highly competent president of the Phipps houses (a 1,610 unit subsidized development in Manhattan): "We thought we could handle ten [problem families], but in retrospect I think it was a mistake . . . we wanted to be helpful. We found out later that we could not work miracles."

These and associated findings seem to indicate that directly subsidized private and semiprivate, as well as public, housing developments with effective and dedicated management and economic viability can successfully house most of the urban poor and near poor, but not all of those with deep seated social and psychological problems. This, of course, should be apparent. Subsidies, by their very nature, are economic instruments; thus their efficacy is primarily in the area of ameliorating the housing deprivations of those whose problems are basically economic. Even in this area, they are often inadequate to deal with the total economic need.

In periods such as the middle 1970s and subsequently, when operating costs were skyrocketing in rental units at all price ranges, the disadvantaged were continuing to experience high rates of unemployment and lagging increases in earnings. Effective demand for housing in this segment of the market, as in all markets, called for steady and adequate incomes. In the case of low-income households, housing subsidies often had to be supplemented by income supplements if collectable rents were to be adequate to carry the development.

The social policies and human resources required to rehabilitate multiproblem families and individuals will seldom be forthcoming in suburbia. Their needs must be met, as the concept of Model Cities recognized, in the locale where most of them live and will continue to live for some time. Involved are drugs, crime, and fear, which, in turn, occasion maintenance, housing management, and community problems of unprecedented magnitude. But as long as there are pockets of poverty and class and color unemployment, only superficial results will be forthcoming. The underlying situation is largely an inescapable heritage and the most ugly manifestation of the black, Puerto Rican, and Chicano experience in this nation.

Berry refers to the issue of ameliorating the heavy social costs incurred when "captive individuals" (that is, people who simply cannot get out of the ghetto) are concentrated without access to the real economy. This group continues to have high unemployment, low educational achievement, and a sense of hopelessness and alienation from society. Attention to causes rather than symptoms demands prime concern be given to factors that exclude individuals from meaningful work within the mainstream of society, for work is a measure of worth in the United States. Programs must be aimed at creating job opportunities, intensifying the quality of educational assistance, and improving health and nutrition so that children will be able to learn. Many such efforts as human capital development must be people directed, but a deliberate strategy of fostering geographic as well as social mobility should not be excluded.

The United States has done little at the federal level consciously to affect urban land use policy and practices; yet federal programs significantly accelerated urban sprawl, inefficient use of urban land, and excessive conversion of farmland. The principal federal programs which have been involved are the mortgage program of the Federal Housing Administration, the federal highway program, and certain provisions of the Federal Tax Code. These activities served to facilitate a large volume of long-term mortgages for single-family houses, made thousands of acres beyond the city limits accessible to urban uses, and created attractive tax advantages of homeownership for a large segment of the American people.

Public action designed to influence land-use policies has been basically at the state level. It is targeted primarily to aid farmers, utilizing favorable tax assessment, low tax rates, tax deferment, various forms of agreements and instruments of tenure and development rights designed to encourage retention of farmland, as well as some efforts to increase the economic viability of agriculture. Recently the Department of Agriculture has worked more closely with states in retaining farmland. One of its most effective tools is to direct water and sewer grants so that the urban growth potentially induced would not undermine farming. These and associated programs have had limited success; they have had least impact in the areas which seem clearly destined to conversion from agricultural and associated uses. Many of them present problems of equity, and those most effective have been quite costly. But taken as a whole, they are significant not so much for their result, but because they represent a break with past sole reliance on local land use planning.

One activity controlled by the state and having a major impact upon urban land use patterns is zoning. The states for the most part have delegated this to local governments where its administration reflects small localities' considerations and interests. Urban land use, which is an area-wide issue, is approached by zoning as a local concern. One consequence is that it often encourages sprawl and wasteful land use. The federal and state courts have recently challenged exclusionary aspects of zoning, and in Massachusetts and New York, state legislation was passed to provide relief. In New York, once the statute was applied, there was agitation for its repeal—an effort that soon succeeded.

In 1961 I proposed in the Housing Act of that year authorization of $50 million to provide federal grants to municipalities to acquire land to be kept undeveloped until local authorities had drafted plans for its use by private builders or public agencies. If I were to revise the proposal today, I would propose grants to the states. Justification for the program declared that "land is the most precious resource of the metropolitan area. The present patterns of haphazard suburban development are contributing to a tragic waste in the use of a vital resource now being consumed at an alarming rate." This proposal was ridiculed by Senator Everett Dirksen who declared that it represented "a new dimension in space. It should have come from the Committee on Aeronautics and Space Science. Instead it has come from the Committee on Banking and Currency." This was all that was needed to kill a proposal which many senators considered a serious threat to the prerogatives of free enterprise. The best we could get was a requested $50 million grant to municipalities for the purchase of land for parks.

By the early 1970s there was a serious discussion of providing a meaningful federal role in land use planning. Two years later the Senate passed a bill that would establish a national land use policy, calling for states to develop "balanced" land use plans, protecting environment and recreational opportunities while providing for economic growth. The Secretary of the Interior would be authorized to make grants to assist states in the development and execution of land use programs, to coordinate land use planning in interstate areas, and to coordinate federal programs which had a land use impact. Within the provisions of this proposal (it was not passed by the House) there was a basic challenge: how do we protect ecological interests and still permit necessary and orderly urban growth? Already we have seen the defenders of suburban exclusiveness, limited growth, and no growth defend their position by espousing a rather recent concern for ecology. Champions of excessive standards of clean air, clean water, and associated ecological attributes can stymie urban growth and inhibit the free movement of excluded individuals and groups. Excessive growth controls are more likely to push developers to the urban fringe than to the cities. And because such controls increase the risk of holding large tracts of land, developers are encouraged to return to the pattern of small subdivisions which the conservationists have long deplored. More serious is the implication of placing the administration of the program in the Department of the Interior.

Taken together, the chapters in this volume delineate that the land use problem is basically one involving prime farmland and land use in cities, suburbia, and exurbia. As pointed out above, to date most public action affecting land use in this nation has been directed primarily at resource-use problems. Its basic focus has been, and is, ecological or physical, frequently oriented toward preservation of farmlands. Social consequences have been largely neglected, and direct action to influence urban development has been largely restricted to zoning, frequently to obtain dubious social objectives.

What has long been desirable and becomes required in contemporary America is a land use policy which is concerned, among other things, with orderly urban development. It would attack sprawl, encourage mass transportation (which we are now doing to a limited degree), encourage more efficient use of urban land, and provide greater opportunities for dispersal of lower-income housing.

Although there is always a logical, and certainly a political, danger in citing foreign experience and suggesting that it is transferable to one's country, despite a different culture, set of institutions, and values, such experience has some relevance in this instance. In his summary of some other nations' approaches to land use policy, Mark Lapping places the matter in proper perspective. He also demonstrates that there can be effective land use policy which deals with the issues of preservation of farmland and efficient and socially responsive urban development. In the Netherlands the policy to preserve farmland is part of a larger national planning thrust which includes distribution of population among large and small cities and rural areas. Urban and rural land policies are both involved: the former seeks to consolidate urban land and halt sprawl, especially between large cities. The instrument to achieve this is a system of land banking or advance acquisition of land for urban use. Sweden, too, combines agricultural and urban land use policies, employing land banking to give direction to urbanization and avoid sprawl. France has recently utilized land banking with a reasonable degree of success. These and associated experience support Professor John Raps's statement:

> The advantages of public land acquisition systems are both numerous and compelling. It would enable the community at large to determine the character of its expanding man-made environmental framework and to create convenient, safe, and attractive neighborhoods. Used to prevent sprawl, it would reduce capital costs and annual operations expenses. It would aid the private building industry by assuring a constant flow of improved and conveniently located sites for construction—preplanned, approved for immediate use, and with no uncertainty as to cost. . . . It might help curb spiraling prices of land, which have become a major factor in raising the cost of housing. It may be the only method that can assure a more balanced social and racial mix in new urban neighborhoods.

The concept of advance acquisition of land for urban development has faced great resistance in the United States. This follows from our long tradition of land tenure, nurtured in an agricultural past and basic to our disposition of federal lands. Equally important is our general attitude toward land—an attitude which Marion Clawson has aptly described as the idea of unrestricted landownership. Other impediments to accepting the proposal are the commitments to local government control over urban land use, the tradition of speculation in land, and the fear of tax losses to local governments.

Yet advance control of *urban* land is not without precedent in this nation. Public ownership was the basis upon which many American cities were developed. Included in this group are Washington, D.C.; Annapolis; Williamsburg, Virginia; Los Angeles; San Francisco; Santa Fe; and Savannah. Much of Manhattan was at one time publicly owned. In its 1937 report, *Our Cities—Their Role in the National Economy*, the National Resources Committee advocated advance acquisition of land for urban use. It urged that loans be made to urban communities to enable them to better control urban development, to combat land speculation, and to have urban land available for low-rent housing, recreational, educational, and other public facilities. The report of the Douglas Commission, the report of the American Institute of Architects, and the report of the American Law Institute also advocated this approach to urban use.

Respect for the national environment and recognition of human needs have increased in recent years in this nation. Increasingly we realize that our productive farmland is not inexhaustible. The formation of new middle-class households in the central city and the growth of exurbia have occasioned a new concern for urban development. Many Americans are asking what is the future for our cities and the suburbs. These developments may converge to create a climate in which an effective land use policy could be forged. As in any major policy decision, there would have to be trade-offs and basic details would have to be worked out. What European experience suggests is not a detailed model, but a possible workable approach modified to fit the peculiar institutions of this nation. European and Asian experience also demonstrates that rural and urban land use patterns interact and thus become self-destructive if not coordinated.

In the final analysis we have to ask ourselves these basic questions: (1) Do we *want* to make sure to preserve enough productive farmland to feed our present and future population and provide a surplus? (2) Can urban development be approached in a comprehensive manner so as consciously to integrate its basic geographic elements: exurbia, suburbia, and central cities? (3) Can and should cities be renewed and, if so, are we prepared to pay the necessary price?

It seems to me that this is a propitious time to pose these issues.

Charles R. Frink and
James G. Horsfall

4

The Farm Problem

"Where ignorance is bliss, 'tis folly to be wise."

So spake Thomas Gray in the middle of the eighteenth century. Americans are ignorant of the desperate loss of prime farmland in the twentieth century, but we doubt if it is "folly to be wise."

The Malthusian Doctrine

During the decade of the 1790s, the Reverend Thomas Malthus worried about Britain's food supply. He could see that the population was expanding geometrically and was gobbling food in proportion. In his day the only route to more food was to expand farmland geometrically. The blue ocean around his "tight little isle" told Malthus, however, that this was impossible, that no more land could be had. What was worse, the cities were flowing out over the land that was there. How could a nation survive if it demanded more food on the one hand and destroyed its farmland on the other?

JAMES G. HORSFALL *is director emeritus and S. W. Johnson Distinguished Scientist at the Connecticut Agricultural Experiment Station, with which he has been associated for more than forty years. Prior to that time, he taught at Cornell University and worked with the New York State Agricultural Experiment Station. Dr. Horsfall is a member of several scientific associations, including the National Academy of Sciences, has served on a number of national and international advisory committees, and has been honored for his scientific work both in the United States and abroad.*

No wonder he came to the lugubrious conclusion that Britain would eat itself into starvation as in truth the Irish did a half century later with the help of a new potato disease which aggravated a singularly damaging land tenure system.

Science Outmodes the Malthusian Doctrine

In the eighteenth century, well ahead of Malthus, another Englishman, Jonathan Swift, threw down the challenge to make two blades of grass grow where only one grew before. A century later, Samuel W. Johnson of Connecticut picked up the challenge by deciding to put science to work for agriculture. Science has been so successful during the third century after Swift that many citizens have forgotten agriculture. They think that their food comes from the supermarket.

The urban and urbane magazine *Newsweek* was so impressed that by 1965 they ran a story about the outmoded Malthusian doctrine. We wonder if the thesis is really outworn. We suspect that neither Egypt nor India would say so. In the United States the population still expands by birth and immigration and the prime farmland contracts. Are we living on borrowed time?

THE RISE OF AGRICULTURAL SCIENCE IN AMERICA

How did science delay the Malthusian moment of truth? In 1798 most seaboard states of America were in very much the same position as Malthus's Britain. Since we know our state of Connecticut best and know that it was typical of most of the rest, we shall discuss it. In 1798 land was "running out"; fertility was declining; prime farmland was all under cultivation; and the population had expanded tenfold in the previous century. Like its neighbors, Connecticut set out to improve its agriculture. It set about to "make two blades of grass grow where only one grew before," in short to improve the ability of its land to produce more food. In 1794 it established an agricultural society to stimulate competition among farmers, and it set out to put the new science to work.

Perhaps we should start with Benjamin Silliman of Yale who, beginning in 1802, taught and practiced some of the earliest chemistry in America. In 1818 he founded a new journal, *The American Journal of Science*. The flyleaf says it would be devoted to "mineralogy, geology, and other branches of natural history, including agriculture." On the basis of his concern for agriculture, Silliman was commissioned by Congress in 1820 to do a treatise on sugar cane growing and sugar manufacturing.

Silliman and especially his son, also Benjamin, enlisted able students in the study of agricultural science. One of these was Samuel W. Johnson who, in the early 1850s, joined the scientific trek to Europe. While working in the chemical laboratory of the great Justus von Liebig, he discovered the German *Landwirtschaftliche Versuchsstation*, literally the "Agricultural Experiment Station." Johnson wrote home that such an institution should be established here to put science to work for agriculture, and that is what he did.

The Persuader Was Fertilizer Analysis—Johnson was a shrewd tactician. He knew that science would have to do something for society if his idea were not to die. Accordingly, he would analyze fertilizer to prevent fraud. Everybody knew that plants need fertilizer, but nobody could distinguish New Haven Harbor mud from Peruvian guano in a bag labeled "fertilizer." Johnson's chemistry could. America was as materialistic then as now, and it could and did appreciate this help from science. It still does.

His first significant success was to be appointed in 1857 as chemist to the Connecticut State Agricultural Society. For them he analyzed fertilizers and fearlessly published the results in terms of their value in dollars and cents. This is the first case of consumer protection in America.

THE FOUNDING OF LAND-GRANT COLLEGES

And then came the Civil War. Congress realized that the war could have been shortened had the North been industrialized. It was impossible, however, to industrialize a country which needed 90 percent of its people to feed itself. The ratio of farmers to nonfarmers had to come down; 10 percent was too small for the labor needed by industry. The additional labor had to come from the farm, but it could not unless farm efficiency were vastly improved.

Therefore Congress moved. It elevated the dignity of agriculture by raising it in the national capital to departmental rank and by creating for it a brand-new system of college education. This became known as the Land-grant College System.

By legislative act Yale became the land-grant college for Connecticut, and Johnson became professor of agricultural chemistry. Johnson's reputation rose rapidly. In 1866 he was elected as the first agriculturist to the National Academy of Sciences when the Academy was only four years old and he was thirty-six.

The war killed the state agricultural society, but it broke the log-jam for tax support for agriculture and the food supply. Accordingly, after the war, Connecticut set up a State Board of Agriculture as an official state agency. This board promptly assigned tax funds to Johnson to analyze fertilizer.

THE LAST ACT

The stage was now set for the last act. The land-grant colleges had been tax supported for eleven years. Johnson was ready to move. In 1875 he persuaded the State Board of Agriculture to promote a legislative bill to establish an agricultural experiment station where theory and practice would march together. The idea grew and flourished. Within thirteen years, fifteen other states had established experiment stations, largely with Johnson's design. Congress encouraged the stations with formula funds provided by the Hatch Act of 1887.

Boom and Bust

And thus was born the idea that government would put science to work for agriculture. The rest is history. The results are writ large for anyone to see. Not only did science make two blades grow where only one did before, it made three blades grow, then four, sometimes five or six. Land productivity rose sharply. Malthus was wrong. A nation could explode its population and survive. It could not make more land, but it could make the land it had more productive.

The time course of potato yield is a fitting illustration because it shows two rate increases (Figure 1). The long lag period at first lasted nearly fifty years. Then basic research showed that the "running out" disease of potatoes is due to a viral infection that is transmitted through the seed tubers. This discovery was transduced into the technology of growing virus-free seed stocks in the Far North. By 1925 yields began to climb as the new technology caught hold and held until about 1945.

Even during this period, however, potatoes were still afflicted with their other classical enemies: flea beetles, leafhoppers, and the foliage blight diseases. The only remedy was Bordeaux mixture, a complex slurry of copper salts. It dwarfed the potatoes, but its pest control more than offset the damage. In 1945 there appeared two new pesticides, DDT and zineb, that controlled the triumvirate of pests. They did better than Bordeaux and without dwarfing the plants. The yields took off like a rocket. They increased more than threefold in the next thirty-five years. In Connecticut they increased four times so that in 1975 Connecticut farmers grew as many potatoes as in 1875 on one-fourth the land. Thus, four tubers grow where only one grew before. Jonathan Swift would like that.

Similar rapid rises in land efficiency could be shown for many crops. Land productivity rose much more rapidly than did the size of the population to be fed. Surpluses appeared and agriculture entered a bust period; we supported farm prices at parity, set up a soil bank, and dumped potatoes in the ocean. Some even proposed that we should plow under every third pig and every third agricultural scientist.

Secretary of Agriculture Henry C. Wallace wrote in 1924, "The increased productive efficiency which normally would have meant prosperity brought bitter fruit instead." Later, his son, Vice President

Fig. 1. Average yield of potatoes in the United States, 1875–1975.

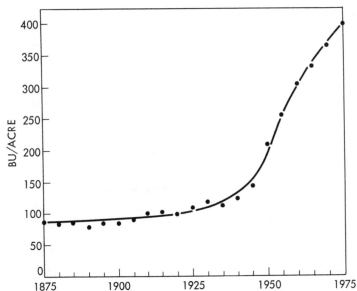

Henry Agard Wallace, argued that the trouble was not too much science in farming but too little science in economics.

LAND PERCEIVED AS EXPENDABLE

All of this reinforced the folklore that in America land is expendable. Thus we acquiesced happily to the rape of farmland by houses, highways, airports, and shopping centers. We were both ignorant and blissful.

Then in 1973 the applecart was upset by the purchase of a few million bushels of wheat by the Russians. Food prices rose sharply. Housewives tried to imitate Canute. They would command the tide to stand still by boycotting the supermarkets. The secretary of agriculture tried in his way to stem the tide by taking all the land out of the soil bank. The tide rolled right over the ladies and the secretary and still rolls. Inflation continues, with food prices often seeming to lead the way.

Other Counterproductive Moves We Make

To cover green farmland with black pavement, gray concrete, and cellar holes is counterproductive enough, but, in our ignorance, we make many other such moves.

The Well Pumped Dry—The old saying is that "you never miss the water until the well runs dry." The wells are running low or dry over millions of acres in the arid Southwest. A visit to the University of Arizona in 1978 revealed that the grass in its beautiful green quadrangle had burned brown. The officials said that the University's water comes from wells where the water table is falling six feet per year. We are told that Arizona policy now is to encourage industry that uses little water, to take the water from farmers, and to quit growing food and to buy it. From whom? We wonder.

Salting the Irrigated Land—The farmers of the Middle East invented irrigation many millenniums ago. This enabled them to build great cities like Ephesus, Nineveh, and Ur, but now Ephesus, Nineveh, and Ur are but mere mounds of rubble because the irrigation water brought in salt that was not later leached out. The land was sterilized, and their cities died. No wonder they cast such greedy eyes on the Egyptians where the annual Nile flood took care of the salting. Pasargadae, the capital city in Persia that served Cyrus the Great, circa 550 B.C., was nestled in a fertile irrigated plain. Essentially no crops grow there now, however, because the land salted up and the city died. Looking now across the sterile plain, you can see only a lonely column or two, silent reminders of his huge audience hall. Cyrus's above ground tomb now stands starkly alone in the man-made desert. Are we living on borrowed time? Do we think that it cannot happen here?

Spewing Air Pollutants—We poison not only ourselves with air pollutants, but we poison our plants as well. For example, oranges and Zinfandel grapes cannot be grown in parts of California, and potato yields in the Connecticut Valley are falling steadily as smog creeps out from nearby Hartford and Springfield. Acid rain from the oxides of sulfur and nitrogen in the air are other potential hazards for our crops.

A Second Malthusian Moment of Truth?

Are we, 200 years after Malthus, approaching a second moment of truth? The problem before us is: can such rapid rises in land efficiency be extended? Are we living on borrowed time? Let us now examine that. The last section of the potato curve in Figure 1 has begun to plateau in many parts of the country. It looks as if we are squeezing potato land for about all it is worth.

The National Academy of Sciences began to examine this question in 1971 and concluded that the productivity of soil for many crops is leveling off. If we divide the yield in pounds per acre for all crops in the nation by the consumption of food in pounds per capita, we obtain the carrying capacity of the land for people as shown in Figure 2. We can feed one person on about one acre. Two centuries ago Malthus sensed the implications of this curve. Are we hurrying on toward a similar rendezvous with destiny?

We must pause and ask ourselves some questions. Are we bumping our heads against a real ceiling of land productivity? Can we raise the ceiling? Has science done all it can? Are we doing the right kind of science? Is our vaunted basic research being transduced into the proper technology?

THEORETICAL AND USEFUL RESEARCH ARE OUT OF STEP

Americans believe that money can buy anything. Are we spending enough money on agricultural research? It is easy to say, "No, we

Fig. 2.   *Carrying capacity of farmland in the United States, 1910–1975.*

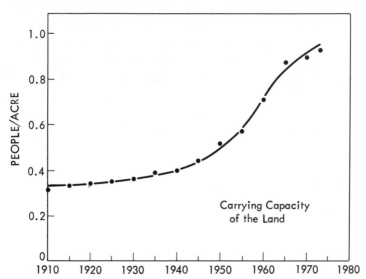

are not." The National Academy of Sciences' *Report on Agricultural Production Efficiency,* published in 1975, says that expenditures have risen since 1950 at the rate of 8.7 percent per year compounded. The legislative bodies have not neglected us altogether, although they have probably not done much more than keep ahead of inflation. We think that something different is needed, not necessarily more money.

For their first seventy-five years the experiment stations followed Samuel Johnson's dictum that theoretical and useful research must march together, each fortifying the other. Basic research in human nutrition gave us the importance of vitamins. Basic research in poultry nutrition gave us cheap chicken meat. Basic research in genetics gave us hybrid corn, hybrid chickens, and disease-resistant plants.

The operator verb here is *gave.* That verb condenses a three-stage operation: (a) basic research to understand nature, (b) useful research to transduce it to technology, and (c) education to transfer the technology to farmers.

At about the seventy-five-year mark in the life of the experiment stations, a subtle change set in. Acting under intense pressure from professors in the "groves of academe," Congress pacified them by setting up an agency to subsidize pure research of their own choosing, provided only that it not be aimed at being useful. Thus, pure research became the stuff of status. Useful research was pushed down in the snob scale. The land-grant colleges, where agriculture had been important, became state universities where it was not. Useful research gained no tenure points for its devotees. Moreover, Johnson's fears that teaching would take precedence over research had been fulfilled: until 1952 about half the station staffs were full-time researchers; but then a decline began, until in 1973 only 31 percent of the scientists at agricultural experiment stations devoted themselves wholly to research.

At this point, agricultural scientists deserted useful field research for laboratory research. Field research became impure research. By 1979 Horsfall was impelled to ask, "Are we smart outside?" and Norman Borlaug, the first Nobel Laureate in agriculture, was forced to say, "Our research must be good, but it must be good for something."

No longer did theoretical research and useful research march together, each fortifying the other. It was the second stage in the process that was neglected. Collectively, agricultural scientists were no longer transducing pure research into "impure" technology. As a result, by the ninetieth year of the stations, many yield curves had reached the point of inflection and were leveling off, as we have seen.

As scientists, we are optimistic that we can once again become smart outside, and hence we conclude by examining the future of our food supply. Since more food requires either more acres or increased production per acre, we look first at our supply of land and second at our supply of technology. Finally, we offer some suggestions for organizing for research to insure that we may eat.

Food for the Future

LAND—"THEY MAKE SO LITTLE OF IT NOWADAYS"

So said Will Rogers long before our present concern for the preservation of agricultural land. Perhaps he foresaw the message of the dust bowl more clearly than our planners in Washington. As late as 1974, they said in effect: don't worry—our computers predict that we will have ample food in the year 2000 and ample land in reserve in the event that we are wrong.

In its study of agricultural production efficiency, the National

Academy of Sciences compared various projections of agricultural productivity to see how they had fared as time had tested their accuracy. Three studies with target dates of 1950, 1960, and 1970–80 consistently underestimated population growth and farm output and overestimated per capita food consumption. All three projections failed to anticipate the increased consumption of meat and decreased consumption of food grains that we now know are inevitable as per capita income increases. More recent projections have been refined to correct these inadequacies, but they failed miserably to anticipate the enormous increase in food exports that has occurred in the last decade.

Of greater concern, projections that we examined rely heavily on returning diverted cropland to production. For years, we were told that we had at least sixty million acres of cropland in reserve, although its whereabouts was not well documented. In 1973 and 1974, all the land in retirement was released to production. By 1976, G. E. Schuh wrote: "To the surprise of many, only thirty-seven million of the sixty million acres in retirement actually came back into production." What's more, the productivity of these idle acres was not nearly as great as that of land already being farmed. The farmers could have told us that—they were too smart to have put their best land in the land bank at the prevailing interest rate.

Realizing that we had used most, if not all, of our reserve cropland, we looked to see where it had gone. Apparently, the exodus to suburbia had consumed enormous amounts of land for roads, houses, shopping plazas, and other amenities of the good life in the country. It was estimated that about a million acres a year had been converted to these so-called higher and better uses. For the first time, we began to realize that the food-people equation of Mr. Malthus was not outmoded and that we had been living on borrowed time.

As if this were not enough, we also discovered that the rate of disappearance of agricultural land was not well known. The estimate of one million acres per year was largely based on 1970 Bureau of Census data. By 1975, some estimated the rate to be as high as two million acres per year. Then came the blockbuster. In a report published in 1977, the Soil Conservation Service (SCS) of the USDA updated their 1967 Conservation Needs Inventory and reported that, between 1967 and 1975, three million acres per year of rural land had been converted to urban uses or covered with water. Even worse, the SCS estimated that for every acre urbanized, another acre was isolated by "leapfrog development." Their final estimate of the rate at which farmland was lost was a shocking five million acres per year.

In order to judge the significance of this rate of disappearance of agricultural land, we need to know the total amount of cropland in the United States. Again, we find substantial discrepancies. The 1977 USDA-SCS study found between 400 and 467 million acres of cropland, reflecting "definitional and procedural differences." Of this uncertain total, 384 million acres were considered prime agricultural land. The total land used for crops in 1974 was reported to be 363 million acres. Although not all prime land is farmed, the comparison of 363 million acres cropped versus 384 million acres of prime land suggests we have a very slim reserve of land.

Since the first warnings of the increasing rate of disappearance of agricultural land came from the urban Northeast, some belittled its importance since this was no longer the breadbasket of the United States. But then we learned that the Corn Belt and the rain-fed Southeast have lost more prime farmland than any other region of this country (Figure 3)! The local newspapers agreed. In July 1979, the *Des Moines Register* ran a six-part series on "Vanishing Acres." Two weeks later, the *Atlanta Journal and Constitution* headlined a

Fig. 3.   Loss of prime farmland in the United States, 1967–1975.

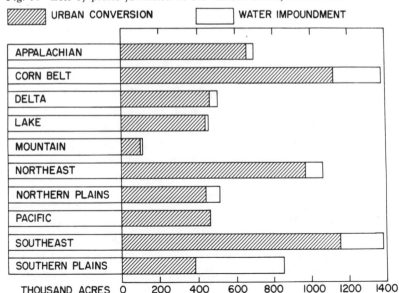

quarter-page story "Urban Sprawl Is Threatening to Destroy Vital
U.S. Natural Resource—Farmland." The option that the Northeast—
or any other region—go industrial and import its food from other
regions is rapidly losing whatever viability it once had.

The American Land Forum has called the USDA-SCS report "as
important a warning to policy makers . . . as this department had
issued since the dust—and depression—ridden 1930s." To their credit,
the USDA has shaken off their earlier complacency and, with the
Council on Environmental Quality, has commissioned a National
Agricultural Lands Study to provide "a more complete understanding of the availability of our agricultural land base." Unfortunately,
the study is off to a slow start since there is considerable disagreement
about the definition of agricultural land. Traditionally, we have used
physical and chemical measures of the capability of the land to produce. Now, we must add the economic criteria of suitability and the
even more elusive social and environmental factors of availability.

To illustrate the problem, consider the dairy farmer in Connecticut
who rents one-half of the land he farms. Typically, the rented land
is in "isolated acres," five to ten miles from the home farm. Economics
dictate against hauling manure to these acres, and manure spilled on
the road along the way virtually guarantees that the neighbors will
call the cops. Thus, the manure accumulates on the home farm. If
the land is level, as in the Connecticut River Valley, nitrates may
appear in the ground water. If the land is hilly, as in the eastern
and western uplands, phosphorus may increase in nearby lakes and
streams. Small wonder that the farmer may decide to sell and put
the money—not the land—in the bank.

While the effort to inventory our cropland is laudable, it falls far
short of the efforts that will be needed to insure our food for the
future. Presently, we obtain all of our food and fiber from about one
acre of cropland per person, with an additional half-acre devoted to
exports of grain in exchange for oil. Thus, while we fret over the
exact rate, we are losing—every year—our ability to feed three to five
million people.

HIS MAJESTY'S LOYAL OPPOSITION

As you might expect, not all agricultural thinkers agree that land
is as important as we do. Those of us who do are relegated by the
others to the limbo of the outmoded Riccardo theory of the "original
and indestructible powers of the soil." In his speech accepting the

Nobel Prize in 1979, Professor Theodore W. Schultz, agricultural economist, has a section "Land is Overrated." "By means of research," he says, "we discover substitutes for cropland." His evidence derives from the logarithmic phase of the productivity curves examined by the National Academy of Sciences, as shown in Figures 1 and 2.

Yes, research did develop technologies that substituted for land. Schultz must have overlooked, however, the recent end of the curves showing that productivity is leveling off. Yes, agricultural science continues to stretch our acres, but the rate is slowing.

THE GEOMETRY OF AGRICULTURE

A curious aspect of biology is that animals move and use chemical energy, while plants stay still and accumulate more energy than they consume. The fact that crop plants are immobile gives agriculture a separate geometry from other industries. Industries improve efficiency by bringing men, materials, and fuel together in a closely knit geometry called mass production. This does not apply to agriculture. Crop plants can be congregated only so much and no more. They must be spread over the ground in a diffuse geometry so that they can capture enough sunlight and carbon dioxide to make a crop. Plants do not lend themselves to mass production techniques, as several business conglomerates are discovering to their chagrin.

Since farm animals can consume energy as food, they can be congregated, and animal agriculture is turning toward concentrated production and away from the extended geometry of crop agriculture.

And so it is clear that the procurement of energy from the sun sets the geometry of agriculture and calls the turn on many of its functions and organizational patterns.

IS HYDROPONICS A VALID OPTION?

Some persons, forgetting the geometry of agriculture, have proposed hydroponics. Yes, you can sometimes squeeze a little more yield per acre from hydroponics than from an equal area of soil agriculture, chiefly because of better nutrition and freedom from root diseases. As a rule of thumb, we can say that food crops are most productive with about four acres of foliage for each acre of land. If five acres are produced, one acre of foliage is shaded out and dies. Hydroponics will not carry a significantly larger leaf area per acre than soil agriculture. The advantage is not enough more to compensate for the necessity of tanks, and pumps, and energy.

USING SALTY WATER

A proposal frequently made, but not yet implemented, is to induce mutations in food crops so that they can grow in brackish water. This is a discouraging proposal because if this were feasible, the ancient farmers of Persia or the "Fertile Crescent" would have picked up such mutations over the years as their fields gradually salted up. Drip irrigation uses vastly less water than surface or overhead irrigation and therefore salts the land more slowly. This has developed technological problems but is moving fast as a possible technology. It will delay the salting process but will not prevent it in the long run.

POSSIBLE BIOLOGICAL BREAKTHROUGHS

Several national study commissions have produced volumes on the subject of biological limitations to food production, and we will not attempt a comprehensive review here. Since we seek a quantum increase in crop production, the odds for success of any particular approach are chancy, but the stakes are obviously enormous in today's hungry world.

Increased photosynthetic efficiency and enhanced biological nitrogen fixation are two items at the top of most lists of research needs. A third, enhanced transport of photosynthate and other metabolites to harvestable portions of the plant, is intimately linked with the first two but has not been viewed in the same glamorous light. We believe that all three offer hope for substantial increases in crop yields. Increased reproductive and feed efficiency in animals offers similar hopes for increased production of animal products, but we shall confine our remarks to crops since they are at the base of the food chain.

Photosynthesis—That yields can be dramatically increased by carbon dioxide fertilization has been amply demonstrated in greenhouses and in closed chambers in the field. Despite the hopes of some, this route to increase yields is not feasible in open fields since almost all of the carbon dioxide added rapidly diffuses away. Nevertheless, these experiments demonstrate that yields can be increased if net photosynthesis can be increased.

We use the term net photosynthesis, since much of the carbon dioxide fixed during daylight hours is released in respiration during the dark. If some of the energy released is not required for essential plant functions, crop productivity could be improved by reducing this dark respiration. However, an even more wasteful process of daylight respiration has been found in many crop species. Israel Zelitch, reviewing this subject in *Chemical and Engineering News* in 1979, points out that in most crops, photorespiration is about 50 percent of net photosynthesis. The potato, whose yield history is shown in Figure 1, is one such crop where the potential exists for another doubling of yield beyond the increases already achieved with fertilizers and disease and insect controls.

A few crop plants have efficient photosynthetic systems, and in these crops, such as corn, sorghum, and sugar cane, photorespiration is less than 10 percent of net photosynthesis. The biochemical explanations for these differences are reasonably well understood, and hence hopes are high that this efficient mechanism can be transferred to other crops by genetic manipulation. In addition, it may be possible to develop chemicals that would slow photorespiration if genetic means are not forthcoming.

Translocation—The migration of photosynthate from the vegetative to the storage structures of plants often controls the amount of grain actually harvested. This is particularly important in soybeans, where increases in yield have been meager in comparison with many other crops. Since soybeans have now surpassed corn as the principal agricultural crop in the United States, the potential impact of increasing soybean yields is considerable.

The rate of photosynthesis in the leaf determines the amount of carbohydrate available and the rate of translocation out of the leaf. The utilization of these materials by various growth or storage organs determines the distribution patterns of translocation within the plant. Since many areas compete for photosynthate, the challenge is to earmark as much as possible for the harvestable portions. Presently, much effort is devoted to increasing our understanding of translocation, and no single strategy for increasing yields has emerged. Since the national average yield per acre for soybeans is about one-quarter that of corn, however, the payoff can be enormous.

Biological Nitrogen Fixation—The ever increasing cost of nitrogen fertilizers makes the prospect of enhanced biological fixation of nitrogen increasingly attractive. Most legumes currently can fix only part of their nitrogen needs and must obtain the rest from soil or fertilizer nitrogen. However, the fraction provided by nitrogen fixation is variable among our legume crop plants. Thus, there is considerable room for improvement of fixation efficiency. Although the impact of improved nitrogen fixation will be less in wealthy countries, the potential increase in world food production is substantial.

The difficulties of inducing nitrogen fixation in nonlegumes are considered greater than those of genetic manipulation of photorespiration. This is so in part because inducing nitrogen fixation requires changing the plant plus finding a strain of symbiotic organisms compatible with the new plant. It may be more profitable to search for organisms living in the soil that can either fix nitrogen presently or could do so if manipulated genetically.

We believe that the odds against this approach are large, and hence it is last on our list. However, we would be delighted to be proven wrong.

ORGANIZING FOR RESEARCH

The discovery in the mid-1970s that the rise in agricultural production efficiency was beginning to show signs of slowing was sufficiently unsettling that agricultural research was suddenly exposed to scrutiny by experts both inside and outside the agricultural establishment. One result was the proliferation of studies to establish research priorities, as we noted previously.

Debate over the quality of research was sparked by a report by the National Academy of Sciences in 1972 that much agricultural research was "outmoded, pedestrian, and inefficient." The paradox that agricultural science was ranked low by academicians yet still undergirds efficient scientific agriculture was examined in detail in subsequent editorial pages of *Science* magazine. The resolution of this paradox appears to us to lie in the principles enunciated by Samuel Johnson a century ago. Agricultural research must be attuned to the needs of its clients, and theory and practice must march hand in hand. To the extent that these principles were neglected in the halcyon days of the fifties, sixties, and seventies, agricultural scientists can indeed be faulted.

The most compelling evidence that theory and practice must be reunited comes from recent economic studies of the returns to society from investments in agricultural research. Beginning in the 1950s with the pioneering studies of Zvi Griliches, who estimated that the annual return on investments in research in hybrid corn were of the order of 35 to 40 percent, a considerable body of information has developed on economic returns on publicly supported research. Selected data for the United States taken from a 1979 report in *Science* by Evenson, Waggoner, and Ruttan are shown in Table 1. Clearly, agricultural research is undervalued as an investment.

TABLE 1. ESTIMATES OF THE RETURN FROM INVESTMENT IN
AGRICULTURAL RESEARCH

Investigator	Year	Commodity	Period	Annual Return (Percent)
Griliches	1958	Hybrid Corn	1940–55	35–40
Griliches	1958	Hybrid Sorghum	1940–57	20
Peterson	1967	Poultry	1915–60	21–25
Schmitz & Schmitz	1970	Tomato Harvester	1958–69	37–46 *
Peterson & Fitzharris	1977	Aggregate	1937–42	50
			1947–52	51
			1957–62	49
			1957–72	34

* Assuming no compensation to displaced workers. Drops to 16–28 percent assuming compensation for 50 percent of earnings loss.

This study also investigated the return on science-oriented and technology-oriented research between 1927 and 1950 and found rates of return of 110 and 95 percent, respectively. However, the study showed that science-oriented research had a high payoff only when it was coupled with technology-oriented research. This the authors called "articulation."

The study also showed that decentralization of research, as between a central state station and substations, also had a positive interaction with technology-oriented research. Thus, contrary to centralizing decisions in Washington, the decision to spread agricultural experiment stations across the country, aided by funds from the Hatch Act, appears to have been a judicious one.

Evenson, Waggoner, and Ruttan conclude that a nation concerned with increasing its productivity by research across the board can learn from agriculture. We know that Malthus predicted that land could not always produce enough food to feed a rising population. Research, for a time, made Malthus look wrong. Now with our diminishing farmland, we begin to wonder if Malthus was right, but we hope that further research will again make him appear wrong.

Robert G. Healy

5

Landscape and Landowner:

Issues of Land Tenure in Rural America

Anyone who has spent much time driving the highways and back roads of rural America cannot fail to be impressed by the tremendous variety of the nation's rural landscapes. In some places vast acreages are devoted to producing a single commodity, whether in the fields of soybeans or corn in Illinois and Iowa or the endless ranks of Douglas firs in the forested Pacific Northwest. Elsewhere, several rural land uses are jumbled together—here a farm, there a roadside business, over there a scrap of forest, perhaps with a dilapidated sign offering "recreation lots, 5+ acres." A significant amount of land seems devoted to no apparent use at all.

This landscape, with all its implications for the supply of natural resources and for the quality of the environment, is for the most part not the product of social or governmental planning or coordination.

ROBERT G. HEALY, *director of the Conservation Foundation's Rural Land Market Project, has, since joining the staff in 1975, conducted research on a variety of land use issues and has directed a major study of coastal land use regulation in California. Formerly on the research staffs at the Urban Institute and Resources for the Future, Dr. Healy has also taught city planning at Harvard University. In addition to writing numerous articles in professional journals and editing, authoring, and contributing to volumes dealing with land use, he is coauthor of* Land Use and the States *(a study of state-level involvement in land use control).*

(This does not include land actually owned by federal, state, and local governments, about one-third of all U.S. land, but heavily concentrated in the western states.) Rather, it has been formed by millions of decisions made over the years—and still being made each day—by the owners of the nation's 1.3 billion acres of privately owned rural land. The land they own produces nearly all of the nation's food, three-quarters of its wood fiber, and a large part of its water, wildlife, and recreation. Yet the landowner's choices are decidedly individual ones, made on the basis of economic opportunities and on his own personal values, skills, and aspirations.

At times, the land itself mirrors very noticeably the pattern of its ownership. For example, the physical boundary between a hayfield and a woodland more often than not corresponds exactly to the border between parcels found in land records at a county courthouse.

In many cases, however, current land uses conceal the reality of owner identities and of owner intentions. In one rural county on the outskirts of Washington, D.C., thousands of acres of corn are planted each year. Yet the land is not owned by farmers, but by real estate speculators, biding their time until sewers become available and zoning is changed. In many forest areas, an apparent abundance of greenery hides the fact that the best timber has been long removed and the land is owned by weekend recreationists. In several coastal areas in California, pastoral landscapes belie the reality that they occupy land long ago subdivided into building lots. Because of its reputation as an inflation hedge, rural land has become a vehicle for investors more interested in capital gains than in commodity production.

This chapter investigates the current structure of rural landownership in the U.S. and describes some notable recent changes in the identity and motivations of rural landowners. It then raises issues that ownership patterns might present both for the use of land and for the distribution of social and economic power. The chapter concludes with some brief comments on how society might begin to evaluate and deal with the land tenure issues raised.

Rural Ownership: Status and Trends

In theory, the ownership of land is a matter of public record. Names of owners at any one time are listed on property tax rolls, while changes in ownership are recorded by the local registrar of deeds. In practice, however, useful information on landownership is exceedingly scarce. This stems from several causes.

For one thing, ownership information is recorded locally, generally in the offices of one of the nation's 3,000 county governments. The size and assessed value of an individual parcel are usually listed, but the information is seldom cumulated to a county total, much less to a total for a state or region. Moreover, the information given is quite limited—one may learn the owner's name and tax address and the year in which he or she acquired the parcel. But in most cases there is no direct way of identifying personal characteristics of the owner nor of identifying how the land is being used. Finally, owners can conceal their true identities, if they wish, through the use of corporations, partnerships, dummy owners, or bank trustees.

Lack of data has historically been a severe impediment to research on rural landownership issues. Many of the ownership issues that I will outline in this chapter are subject to a wide range of opinion, with a final resolution simply not possible until important data gaps have been filled. As it is, I have relied on published and unpublished U.S. government data, small area studies done by other researchers, and the results of land market studies undertaken in rural areas of six states as part of the Conservation Foundation's Rural Land Market Project.

OWNERSHIP STATUS

According to a survey taken in 1978 by the U.S. Department of Agriculture, the biggest owners of land in the U.S. are farmers, who own more than half a billion acres—38 percent of all privately held acreage. This is certainly not surprising, given the large amount of U.S. land that is devoted to agriculture. Nevertheless, it is interesting to note that 44 percent of farm and ranchland is owned by nonfarmers.

Next in importance are retirees, who hold 190 million acres, or 14 percent of all private land. Many of these are probably retired farmers who either have leased their fields to tenants and continue to live on the farmstead or who have moved off the farm entirely to spend their later years elsewhere.

Corporations own 142 million acres, 11 percent of the total. At least 68 million of these acres are commercial forestland owned by wood-producing companies, most of it in Maine, the South, and the

Pacific Northwest. Also in this category are the rather extensive holdings of mining and petroleum companies, railroads, and agricultural corporations.

The remaining owners are a diverse group—nonfarm rural families, urbanites who have inherited or purchased land in the country, and various types of partnerships and syndicates. More than 10 million acres are owned by real estate dealers themselves. We simply do not know what proportion of owners actually live on their land, although the government estimates that 80 percent of all land is owned by someone resident within the county containing the land and that 94 percent is owned by people residing within the same state.

Foreign ownership of U.S. land has received a great deal of attention in the press, usually in the wake of a handful of large, identifiable purchases. The best available data indicate that, nationwide, foreign residents own less than 1 percent of all privately owned land. The overwhelming majority are Western Europeans or Canadians—very few are residents of oil-exporting nations.

Simply inspecting these major classes of landowners points up some potential differences among them. Farmers are quite clearly interested in using the land, from year to year, to produce commodities. Urban-based owners are more likely to be concerned primarily in recreation, in future rural homesites, or in capital gains. Foreigners, in some cases, may look on U.S. land as a good way to protect capital from inflation, expropriation, or home-country tax collectors.

But there are similarities among the classes as well. Even the most dedicated farmer is quite aware of his land's potential for price appreciation—so much so that land is often jestingly referred to as "a farmer's last cash crop." Urban and foreign investors may not actively participate personally in farming, on the other hand, but they frequently try to generate income by leasing the land to someone who will. And all classes of owners are affected by such macroeconomic forces as interest rates, availability of mortgage money, and anticipated rates of inflation.

MARKET TRENDS

The rural land market in the U.S. has been characterized in recent years by three major trends: a broadening of the motives for rural land purchase, rapidly rising prices for all types of rural land, and changes in the sizes of parcels. The trends have their roots as far back as the beginning of the post-World War II period, but they have accelerated greatly since the late 1960s.

Changing Motives—Rural land has always been purchased for a variety of motives. Farmers and timber companies buy land to use; recreationists look for sites for second homes; speculators seek to anticipate future urbanization. Although supporting data are far from complete, it appears that certain motives for land purchase have been growing in importance over time. One is the desire to own rural land as an inflation hedge. Since 1950, the general level of U.S. prices has risen at a compound rate of 3.8 percent per year. Since 1970, spurred first by Vietnam War budget deficits and later by increases in energy prices, the inflation rate has averaged 6.6 percent per year. In response, investors have tried to place their funds in various commodities which they expect will offer total returns at least equal to the rate of inflation. Precious metals, art objects, houses—and rural land—have been among their favorites. In many ways, this inflation hedging demand has resulted in a self-fulfilling prophecy, for it has contributed to rising land prices and hence made land appear all the more effective in this role.

Increases in farm and forest product prices have contributed to the demand for land suitable for production. Crop prices remained rather low through the 1950s and 1960s but shot upward after a major 1973 Soviet wheat purchase. Strong export markets and the federal government's willingness to raise crop support levels over time have led to continued optimism regarding the long-term crop price outlook. (The 1980 partial embargo on grain exports to the Soviet Union could easily affect this enthusiasm, however.) Timber prices have risen quite rapidly in recent years and have been accompanied by increased demand for species and sizes of trees once thought of little or no commercial value.

The construction of the Interstate Highway System, which began in 1956, dramatically increased access to many rural areas. It expanded the commuting fields of major metropolitan regions, encouraged the location of factories in rural areas, and made many recreational destinations newly accessible to weekend travelers. These factors provided urban residents with new motives for investing in rural land.

Beginning about 1970, a wide range of rural areas began to experience a revival of population growth. This phenomenon not only increased the demand for rural land for immediate use, but it also increased purchases by people seeking to secure homesites in advance of future retirement. A boom in heavily promoted recreational developments, concentrated in 1969–73, led to sales of rural properties ranging from undeveloped rural lots to luxury resort condominiums. These promotions may also have resulted in increased buyer interest in rural land in general.

Finally, the decline in the value of the dollar relative to some foreign currencies quickened foreign interest in U.S. real estate, with productive farmland perceived as extremely cheap by world standards.

Rising Prices—The new motivations for rural land purchase have confronted a rather inelastic supply of rural properties. Long-time rural landowners often proved reluctant to sell, regardless of the price offered. Moreover, the traditional owners themselves did not take long to learn of land's attractiveness as an inflation hedge. Holding onto their property, it seemed, was not only convenient, but also profitable. The interaction of increasing demand and static or declining supply was obvious—steeply rising prices for rural land of virtually all types, virtually everywhere in the country.

Best documented is the increase in the price of farmland. Between 1970 and 1979, the average price of farmland tripled. Comparing 1950 and 1979, the average price rose ninefold. (During 1950–79, by contrast, the general price level rose 320 percent.) Every part of the country participated in the long price boom, although the more heavily urbanized states did rather better than average before 1970, while the Corn Belt showed the greatest increases after 1973.

Timberland prices have probably risen at least as rapidly as farmland prices, although documentation is scarce, A New Hampshire researcher reported, for example, that "common talk among foresters is that forestland prices have gone from $10 an acre to nearly $100 an acre in a decade." A U.S. Forest Service land buyer noted that low-grade timberland in Arkansas, which sold for $12 to $18 an acre in the early 1960s, sold for $50 by 1970 and by 1976 sold for $150 an acre or more. A study of Vermont woodlands found that prices more than doubled between 1968 and 1973, although they then leveled off.

Values of recreational land, as measured by prices paid by the U.S. government for additions to national parks and national forests, have shown substantial increases over time.

Consolidating and Parceling—The changing composition of demand for rural land has also modified the size of parcel in which land is held. In some parts of the country, scale economies in farming have led to land consolidation; elsewhere demand has been strongest for small acreage parcels, causing landowners and land dealers to respond by creating lot splits.

In the more fertile farming areas of the country, the predominant trend has been toward expansion of the size of farms. Between 1950 and 1974, the average size of a U.S. farm more than doubled, reaching 440 acres per farm. Even as small farmers were leaving the countryside, their neighbors were buying the land to add to their own expanding operations.

In other parts of the country, other rural land—particularly woodland and marginal cropland—was being divided. A relatively small amount was cut up into small lots for recreational purposes. A larger, but unquantifiable, amount was being divided into the five- to forty-acre parcels newly in demand by urban investors and seekers of homesites. Investigations by the Conservation Foundation of land markets in selected rural counties found evidence of considerable land division between 1954 and 1976 in all but one of the six counties examined around the country. The only exception was Douglas County, Illinois, where a strong demand for land by farmers dominated the land market. In each case, the division of land was accompanied by a significant increase in ownership of rural land by people whose mailing address was nonlocal.

Ownership Issues

Landownership patterns raise two types of issues. Aspects of each have been the subject of great attention in various political forums and in the press, generally in a piecemeal fashion that has made it difficult to assess their true importance. First is the relationship between ownership and use. Simply put, how do owner characteristics influence choices about land use, particularly those choices involving environmental damage or those affecting the long-term availability of the land for commodity production? For example, is an Illinois farmer, the third generation of his family to own his ground, more or less likely to keep the land in agriculture than is the owner of the neighboring land, a Chicago dentist who has recently bought it as an investment?

A second set of ownership issues involves sociopolitical questions of the concentration of control over wealth and the fruits of wealth. For example, even if a foreign owner makes land use decisions identical to those made by a long-time local farmer, some people question the consequences of foreign control over a significant proportion of U.S. farmland. Other similar issues involve corporate ownership, declining ownership by minorities, and problems of landless young people who want to enter farming.

OWNERSHIP AND LAND USE

In another chapter, Professor Harriss discusses how the land market allocates land among competing uses and points out, in a theoretical way, various institutional imperfections that may cause the land market to send misleading price signals to landowners. Here we investigate the relationship of current ownership patterns to land use decisions, emphasizing situations in which private and social interests appear to diverge.

Ownership and the Speed of Adjustment—The identity of the land-owner probably has a greater impact on the speed of adjustment to changing land use opportunities than it does to the eventual outcome. In the short run, at least, land use choices depend not only on objective opportunities for profit but also on the needs, skills, and personal values of the landowner. When a new opportunity for profit meshes easily with the landowner's needs, skills, and values, land use changes can be swift. For example, the large 1973 Soviet purchase of U.S. wheat made farmers keenly aware of the potential for selling their products, particularly feed grains, in overseas markets. In response, land that had long been devoted to pasture was very rapidly planted in corn, wheat, and soybeans. The number of acres of cropland harvested rose from 294 million in 1972 to 343 million in 1977. The price of land suitable for raising these crops also adjusted, although with a slight lag, to the new opportunity.

But new profit opportunities do not produce such a speedy response when they require heavy investment, greatly changed scales of operation, new knowledge or operator skills, or a change in the landowner's life style. For example, changes in agricultural technology and increases in off-farm job opportunities in the early part of this century made it increasingly uneconomical to operate small farms in New England, the Appalachians, and the Southeast. Nevertheless, the abandonment of these operations was a slow and painful process, extending over decades, even generations. The small farmer was often bewildered and angry, as crop prices were driven down by the competition of larger, more mechanized farms in other parts of the country. Academic experts counseled that the market was simply signaling farmers to "get big or get out." Yet for the individual farmer—and for farm-dependent communities—the adjustment costs were high and the adjustment process a slow one.

In much the same way, many farmers on the edge of expanding cities occupy land whose value for development is many times its value in even the most efficient crop production. Yet they may not sell immediately, for a sale would require that they move off the land and, most likely, permanently sever their tie to agriculture. Their land will in most cases be eventually sold for development, but only after the present operator dies or retires.

The importance of owner characteristics in determining the supply of land is borne out by the results of a survey of farmland owners in three Vermont counties which found that health and age were the two most frequently cited reasons for selling land. Ranking only third was a reference to profit opportunities: "Received a good offer for the land." Other studies have found ownership related to the supply of land for conversion at the urban-rural fringe. One researcher noted that:

> For farmers in the fringe, the fact that a specific parcel of land is irreproducible in supply has important consequences. Farmers cannot reproduce the land parcels they offer up to the market for sale. Their decision to sell may require not only a change in location but the sacrifice of a desirable life style as well.

The existence of ownership-related adjustment difficulties has created some significant political issues. Owners have sought government intervention to insulate them from the effects of market forces, either to permanently preserve their life style or, more frequently, to enable them to adjust the timing of their eventual land sale to their own personal needs. For example, a major stated goal of farm commodity programs has been to maintain the viability of the small, family-owned farm. (Ironically, the fact that most programs support the price of crops, rather than the incomes of farmers, has meant that the larger the farm, the greater the subsidy received by the farmer.) Similarly, farmers have successfully petitioned for changes in tax laws that permit major reductions in inheritance tax on land that is farmed with the active participation of the heir. This change is intended to allow land to stay under the control of farm families and to lessen the role of inheritance tax liability as a major force in bringing rural land to the market.

Examples abound of urban-fringe farms trying to hold out against surrounding development. Most eventually succumb to the pressures of high property taxes, vandalism, and the objections of surrounding residents to the noise, dust, and odors connected with agricultural practices. Here too, owners who do not wish to sell have asked for government help, principally in the form of preferential assessment for property tax purposes.

The Role of the Speculator—The frequency of adjustment difficulties on the edge of cities has created a role for a special kind of landowner—the land speculator. The speculator tries to anticipate future land use changes and to hold a stock of land that can, for an appropriate price, be quickly converted to the newly profitable use. Thus, a speculator might buy a farm from a retiring farmer even though it was unlikely that the property would actually be urbanized for a decade or more.

A major land use issue is whether the amount of land being "ripened" for development by speculators is socially optimal. If speculators guess right about future urban needs, they have performed a useful service in easing land's adjustment from one use to another. If they are overly optimistic about development prospects, however, they can create great rings of unused land around our cities, land withdrawn from agriculture but perhaps never needed for urban expansion. It has also been charged that speculation can become an end in itself as fringe area land is traded from one optimistic speculator to another, at ever increasing prices. Such situations, it is alleged, severely restrict the supply of land available for development and force homebuilders into inefficient patterns of leapfrog sprawl.

The speculative potential of recreational lots has caused what is apparently a vast oversupply of such properties in many scenic areas. It has been estimated that there are well over ten million vacant lots in recreational subdivisions—at current rates of second-home construction, this would be more than a fifty-year supply. The existence of a speculative demand for such properties, principally by woefully underinformed small investors, has resulted in a tremendous amount of needless environmental damage.

Current record-high prices for farms and forestland clearly contain a large element of speculative demand, for such land almost invariably sells at prices far higher than can be justified by current levels of crop or timber income. In this case, however, the speculation is not about future changes in use but rather concerns the outlook for future commodity price increases. I see no evidence that high land prices have, thus far, caused undesirable changes in land use. In fact, they may actually encourage owners to manage their land more intelligently and aggressively in order to generate enough income to offset the high costs of holding land.

Ownership and Parcel Size—Changing parcel sizes in rural areas has some complex, but potentially important, implications for land use. Considering cropland first, the apparent economic advantage of the large-scale farm has been a major factor in allowing a rapidly declining farm population to produce ever increasing quantities of food. Yet large, monoculture-based agricultural operations have been criticized for producing a visually monotonous landscape, for increasing the danger of catastrophic outbreaks of insects and plant diseases, for reducing the rural population available to support small-town businesses, for destroying wildlife habitats, and for being slow to adjust to changing market conditions. University of Illinois researchers, for example, point out that farm expansion in Illinois has created a "grain desert" which has "drained the rural landscape of other economic activities."

The small, rural, residential parcel (or part-time "hobby farm") has also been both praised and criticized. Fields that are allowed to revert to brush are inefficient from the standpoint of commodity production but have, over the years, been welcome additions to wildlife habitats. Small, part-time agricultural operations may not be very productive in an economic sense, but they increase landscape diversity and allow people to enjoy rural living without having to become subsistence farmers. Urban small landholders, even when just weekend visitors, have probably also had a beneficial effect on some rural towns, bringing in new purchasing power to support rural businesses and new attitudes about environmental quality.

The most important issue connected with changing parcel sizes is probably the irreversibility (at least in the economic sense) of parcel splitting. To date, much of the land purchased by urbanites for residential or recreational purposes has been marginal land not currently in much demand for either agricultural or forestry purposes. Although the new owners tend to have neither the knowledge nor the intention of using the property for intensive commodity production, it is reasonable to predict that future increases in food or timber prices could cause them to change their minds. A strong force preventing such a change would be small parcels and scattered ownership.

In the present real estate market the size of parcel considered ideal for recreational or residential purposes bears no relation to the size that is economic for producing food or wood. In fact, there seems to be a growing divergence between the two—increased mechanization makes large-scale production profitable, while high land prices drive down the amount of land that the typical recreationist can afford to buy. Rural parcels, once split and sold to urban owners, may prove very difficult to recombine even if resource prices are considerably higher than present levels. Experience with central city properties during the 1950–70 period is suggestive. Although small downtown lots became economically obsolete, their value did not fall sufficiently to induce private redevelopment and the consolidation of parcels. Rather, such lots remained vacant for decades, striking reminders of how fragmented ownership slows the land market's response to changing conditions.

Ironically, otherwise progressive government land use regulations sometimes contribute to parceling. For example, many counties have zoned their rural portions for minimum building lot sizes of five or ten acres. This requirement may have the inadvertent effect of causing a proliferation of such medium-sized parcels, which are unnecessarily large for residential purposes, yet far too small for most agricultural uses. Similarly, Vermont's state land use law (Act 250) exempts parcels of greater than ten acres from certain regulations. It is said that subdividers have chosen to avoid the costs of compliance with the law by creating parcels just slightly larger than the ten-acre limit.

Until very recently, virtually no data were available to indicate either the size of rural landholdings or how those holdings were changing over time. Table 1 presents preliminary data from U.S.D.A.'s national landownership survey. It shows that the overwhelming majority of private land acreage in the U.S. appears to be held in ownership units large enough to permit efficient commodity production. For example, 94 percent of cropland is held in units of fifty acres or more. And for rangeland, where much larger management units are appro-

TABLE 1.   LAND USE IN 1977 BY SIZE OF HOLDINGS, UNITED STATES
(EXCLUDING ALASKA) *

Size of Holdings **				Land Use			
	Crop-land	Pas-ture	Range	Forest	Other †	Urban & Water ‡	Total
(Acres)				(Percent)			
Less than 10	1.28	2.59	.87	3.12	4.69	32.82	3.32
10 to 49	4.45	9.47	1.59	8.53	9.73	11.84	5.82
50 to 69	2.25	4.57	.45	3.70	3.39	3.23	2.49
70 to 99	5.41	9.41	1.14	6.16	8.05	4.03	5.00
100 to 139	6.19	9.07	.92	6.69	7.25	2.96	5.20
140 to 179	10.77	9.14	2.21	5.65	5.28	3.84	6.67
180 to 259	10.87	12.21	2.39	7.79	5.97	4.55	7.65
260 to 499	20.64	16.71	7.64	10.85	10.80	8.71	13.64
500 to 999	16.88	11.71	11.00	8.08	8.96	6.38	11.91
1,000 to 1,999	10.20	6.06	12.75	4.87	6.47	4.80	8.69
2,000 and over	11.07	9.06	59.04	34.56	29.40	16.86	29.63
TOTAL	100.00	100.00	100.00	100.00	100.00	100.00	100.00

* Preliminary data provided by Linda K. Lee, Oklahoma State University, subject to revision.

** Acres owned in landowners' home county.

† Includes farmsteads, other land in farms, strip mines, quarries, gravel pits, burrow pits, barren land, and all other land not defined elsewhere, including greenbelts and large unwooded parks.

‡ Includes urban and built-up land, transportation uses, water, and miscellaneous land uses for which limited Soil Conservation Service data are available.

priate, more than 90 percent of the acreage is in units of 260 acres or more. Even if these units prove too small for efficient operation (and certainly a fifty-acre farm or 260 acres of rangeland is not regarded as optimum in most parts of the country), the ownerships seem large enough to make consolidation into still larger units feasible.

If the data do not give us cause for immediate alarm, however, they still point to some areas that bear watching. For example, they indicate that 21 percent of private forestland is held in ownership units of less than 100 acres. Foresters disagree on what constitutes the minimum size of parcel for efficient timber management—in fact, they simply have not devoted much attention to the subject. But 100 acres certainly is small enough to make profitable management at least questionable. A combination of small parcels and owners uninterested in forest management implies that a very large amount of forestland may not be available for future wood production. Similarly, the fact that somewhat less than 6 percent of all cropland is held in parcels below fifty acres becomes more striking when one recognizes that well over 20 million acres of land are involved. Considerable attention has recently been focused on the urbanization of farmland, virtually none on acreage whose production potential may have been severely limited by parceling.

Tenure and Land Stewardship—Even when land is currently being used for resource production, the identity of the owner may influence how much attention is paid to controlling erosion and maintaining soil fertility. In 1978, some 282 million acres of farmland, about a third of all farmland, was rented rather than owned by its operator. Many observers contend that leased land is frequently abused by its operator, particularly when the owner lives far away and the lease is only for a year at a time. As a Maryland farmer put it:

> I can go by any farm along any road in this county and tell you if it's owner operated or tenant operated. . . . If it isn't yours, you put nitrogen on real heavy—you get the return right away—but you cut corners on the potash and the phosphorous.

The relationship between land tenure and land stewardship has not been well explored. One recent study in western Iowa does support the "careless tenant" hypothesis—it found that renters were losing 20.9 tons per acre of topsoil annually to erosion, while owner-operators were losing only 15.6 tons. If additional studies bear this conclusion out, it would tend to confirm the wisdom of what is said to be an Appalachian proverb: "The best fertilizer for the land is the footsteps of the owners."

Public Access—Land tenure has also been held to influence attitudes toward permitting public access to rural land. In many parts of the country, private lands supply most of the public opportunity for hunting, fishing, snowmobiling, and cross-country skiing. For example, nationwide, 67 percent of all days of hunting activity take place on private land, with the figure reaching 87 percent in the New England states.

Traditionally, local landowners have allowed one another to hunt or fish on their property. But this is changing as rural land is being bought by new types of owners who have no ties to local people and who may have a personal dislike for such activities as hunting or snowmobiling. Posting of rural land has become widespread. Small parcels compound the problem, for they make it very difficult for a conscientious recreational user to seek permission from owners to cross their land. This is likely to lead to increased trespass and to give landowners an additional reason to post their land against recreational use.

Issues of Resource Control

Objection to the concentration of landed wealth is an ancient political theme; indeed, even the history of Athens and other Greek city-states was marked by recurrent struggles between the landed and the landless. Although its importance is not what it was in preindustrial times, land continues to confer wealth, social status, and political and economic power on its owner. During most of the 1970s a great deal of press attention was devoted to issues of rural resource control, and a number of advocacy organizations emerged to expose concentrated ownership and its abuses. Current issues of resource control include: (a) absentee ownership of rural land, particularly by corporations or by foreign interests; (b) declining ownership of land by minorities, particularly southern blacks; and (c) difficulties experienced by landless young people who wish to become farmers.

ABSENTEE CONTROL

It is frequently charged that ownership of an appreciable fraction of an area's land by absentee owners leads to neglect of important local interests. An absentee owner is assumed to be more likely to bank, buy supplies, and spend money in the city, not at local businesses. Often cited in this regard is Walter Goldschmidt's 1947 study of two rural California towns, carefully matched except that one was surrounded by corporate farms and the other by small, family operations. Although he found the total volume of agricultural production was the same in both areas, Goldschmidt concluded that family farms generally supported "a healthier rural community" with "more institutions for democratic decision making . . . a measurable higher level of living [and] better community (and physical) facilities." His findings were replicated in subsequent studies by others of a large number of towns in California and of rural counties in Alabama.

Large absentee owners, particularly when they are corporations, have also been charged with exercising an unwelcome domination of the local political system. As one southern Illinois citizen group put it, in pointing to the holdings of corporate coal interests, "We believe that whoever controls the land will ultimately control the people." The case of Appalachia, with its history of both outside ownership and economic backwardness, is frequently given as proof of that contention. Indeed, several of the groups most concerned with land-ownership questions have their roots in Appalachia. For example, it is charged that landowner interests there have controlled state and local politics so as to favor lax regulation of mining practices and unfairly low property tax assessments of mineral-bearing and timber lands.

Elsewhere, absentee corporate landowners are accused of using their political power to garner government subsidies, such as cheap water from irrigation projects (a major issue in California's Central Valley) and crop price supports. Corporate owners are also charged with using their local dominance to bargain down prices of raw materials, such as pulpwood, while using their national economic power to drive up prices of their own processed commodities.

Foreign ownership is viewed with particular suspicion—it combines all of the issues of absenteeism with an added dash of nationalism. Because many foreign purchases are for investment, rather than for some obvious agricultural or industrial purpose, they tend to have a disturbing air of mystery. It is perhaps for this reason that the relatively small, albeit perhaps increasing, amount of foreign farm-land purchases has received so much attention in the press and even in Congress. Resentment of OPEC may also play a part, although evidence to date indicates that the foreign purchaser is much less likely to be an Arab sheik than a Canadian nervous about Quebec separatism or a European investor awash in cheap dollars.

Some Americans recoil from the idea of *any* foreign ownership of farmland. Others say that it is tolerable only in small quantities; one writer argues that "foreign investment is like foreigners: a sprinkling does not hurt and will often be beneficial, but large concentrations may be disturbing to national life." Still others agree with the former Department of State official who observed that it is far better for even potentially hostile foreigners to own fixed (and hence expropriable) assets within the U.S., such as land, than it is for them to continue to hold huge quantities of U.S. currency and government

securities which can be rapidly shifted in ways that might threaten
the nation's monetary system.

A variation of the absentee owner issue decries the separation of
ownership from management, even if the owner is a local business-
man rather than a far-off corporation. For example, one Nebraska
research group, noting the increasing scale of farming in the state
and the increased use of professional farm managers, observes that
if current trends continue "it is possible to forecast the development
of a Corn Belt class structure rigidly defined along owner-manager-
worker lines." University of Illinois researchers speculate that one
reason for the growth of the state's "grain desert" is that absentee
landowners find that leasing land to grain farmers is a profitable
way to manage land with a minimum of personal supervision.

ACCESS TO LAND

Although concern over concentration of control over land has
focused on absentee ownership, an equally significant phenomenon
is the increasing scale, both in acres and capitalization, of the "family
farm." In 1954, the average U.S. farm contained 242 acres and was
valued at $20,405. By 1974, the average farm was 440 acres and was
valued at $147,838. In the most productive farming areas in the Corn
Belt and in California, capital values of $500,000+ are the norm.
This quantum jump, which has been compounded since 1974 by
steep rises in land values, means that it is difficult or impossible for
landless young people to enter farming. It is often wryly noted in
farming areas that the only way a young person can get a piece of
land is to "inherit it or marry it."

Several studies have shown that current land market forces heavily
favor the already landed in the competition for new properties. Past
increases in land prices usually mean that the person who already
owns land has a great deal of available equity; that is, his mortgage
debt, if any, is very small relative to the land's current market value.
He can borrow against this equity to buy additional land even though
the price of the new land may be far higher than its current production could justify. Moreover, the expanding farmer is often able
to realize scale economies, spreading the fixed cost of machinery and
structures over an ever greater number of acres. The beginning farmer, on the other hand, is likely to be both a smaller operator and
less able to borrow on equity. Small wonder that in recent years the

majority of farm sales have not been of complete farms, but rather pieces of land added to existing farms.

By traditional definition many of the country's large, capital-intensive farms are still "family farms"; that is, they are operated primarily by the labor of the farmer and his family. One observer of the current farm scene notes ironically, however, that "the greatest threat to the small family farm may very well be the larger family farm—intent on expansion and doing so from a solid financial base."

Minority Ownership—Blacks and Hispanics make up 17 percent of the nation's population, yet they own only 1.7 percent of the nation's land. A number of advocacy groups have focused on this disparity, pointing out that these minorities are thereby prevented from sharing in land's traditional role as a source of economic and political power. Particular attention has been paid to the case of southern blacks, whose control of rural land has fallen significantly during this century. In 1910, black farmers owned over fifteen million acres in the South; by 1970, they owned only six million.

This loss of land has in many cases been the accompaniment of individual betterment, as black owners migrated toward the higher incomes and educational opportunities available in northern and southern cities. Unfortunately, by leaving the land, black owners have missed sharing in the enormous capital gains that have accrued to rural landowners during the last three decades.

Those minority owners who remain on the land tend to face the classic problems of small scale and lack of capital for expansion. In many parts of the South, present owners also have clouded title, for the occupant may share legal ownership with a large number of relatives scattered around the country. One estimate puts the amount of such "heirs property" at one-third of all land owned by blacks in the rural South and notes that such land cannot be bought, sold, or used by the occupant as collateral for housing or agricultural expansion.

A Perspective on Rural Land Tenure

Land tenure is one of the most potent forces molding the rural landscape. Yet as a society, we know very little about who owns rural land and why they own it. Policy makers and researchers alike have begun to suspect that the old structure of rural landownership is changing, under pressure from both the new economics of agriculture and new urban-based demands for land. But the statistical base necessary to describe these trends and to diagnose potential problems connected with them has simply not been available. Very recently, new federal data sources are beginning to appear. They are likely to reassure us that some problems are not so great as we once feared, but they are also likely to raise additional issues.

Only when we have a solid picture of rural tenure issues will we be able to face a second problem—some important contradictions in our land tenure goals. On one hand, we continue to prize the Jeffersonian ideal that "the small landholders are the most precious part of the state." Yet at the same time, we have created an agricultural system that produces food most cheaply on production units that have risen steadily in size and capital value. Another set of contradictory goals is found in our attitude toward the proper goals of landownership. We decry land speculation and the accumulation of land by corporations and foreigners; yet we provide heavy tax subsidies to land investment through the deductibility of mortgage interest and preferential treatment of capital gains.

After we have rationalized these goals—if that is ever possible—we must confront the problem of institutions. The institutional framework within which landowners make decisions has remained virtually unchanged for decades, even as the identity and intentions of the owners have undergone dramatic change. Even now, a few local groups are experimenting with new ways of organizing the ownership and control of rural land, among them land trusts that return land value increments to the community; arrangements for pooling the use of small, scattered parcels; and new, less land-wasting forms of rural settlement. It is not too early to begin developing these new organizational structures.

Frederick E. Smith

6

The Environment

Protecting Natural Values

We value nature in many ways, including aesthetics, recreation, resource management, and conservation. Protecting these values has become an important process in planning at the local, regional, and national level. Urbanization, particularly, is often viewed as a destroyer of nature. Urban people, however, have views of nature and of rural environments that are increasingly romanticized as their own lives are more thoroughly played out inside the cities. For the first time in the history of America, a majority of young urban adults have not grown up in the country.

THE RURAL LANDSCAPE

Human activities result in three general categories of land: urban, rural, and wild. Urban land is dominated by built structures and

FREDERICK E. SMITH *is professor of Advanced Environmental Studies in Resources and Ecology at Harvard University. Previously he was professor in the Department of Zoology before becoming a chairman of the Department of Wildlife and Fisheries at the University of Michigan. President of the Ecological Society of America from 1973–74, Dr. Smith became chairman of the board of the Institute of Ecology in 1977. He has written extensively in the fields of theoretical ecology and applied ecology.*

human activity. Cities usually include a variety of natural elements—shade trees, parks, gardens, urban wildlife—and sometimes contain isolated patches of natural reserves, but in general the human systems dominate and the self-organizing functions of natural ecosystems are fragmented.

Wild land, or wilderness, is land where natural ecosystems flourish undisturbed by human influence. In an absolute sense, most wilderness has vanished from the planet and all has vanished from America, but by stretching the definition, a lot of land is still called wilderness. It may contain trails and campsites, but all human presence is constrained so as to minimize impacts on nature. People trespass as passive observers.

Rural lands are those of agriculture and forestry, where natural systems are modified and put to work producing useful products. Most of the land in the United States and in the world is rural in this sense.

Wilderness (we hope) changes slowly, if at all. Cities have evolved more and more rapidly with time. Rural areas also have changed, especially in the last fifty years.

The traditional rural landscape was a mosaic of fields, pastures, woodlots, and farmsteads. Interspersed were ponds, streams, and patches of natural vegetation. Hedgerows bordered the fields, which were small enough to be worked over in a day or two using horses. The scene, of course, was not natural, but man-made. Yet wildlife was abundant around the borders of agricultural activity, and a lot of nature persisted in this traditional rural landscape.

In the eastern United States this old-fashioned style of family farming probably enriched the mix of natural elements by opening up the forest with pastures. In the central United States the reverse process of planting trees and shrubs around the farmsteads and planting rows of trees as windbreaks also probably enriched the mix of plants and animals.

This picture postcard landscape began to vanish shortly after World War II, both in America and Europe. It supported a kind of agriculture that could not possibly have supported today's populations. The agriculture of the late 1970s was more productive; it depended less upon renewable resource management and more upon input-output management. Part of the process was profound changes in the rural landscape, allowing an increased efficiency in the use of land. Where this was not feasible, as in the northeastern United States, agriculture declined, and the rural landscape slowly reverted to wild or became urbanized.

The rural landscape by the early 1980s was primarily one of intensive agriculture. Especially in the major crop-producing areas, fields were large, hedgerows were scarce, and leftover patches had disappeared. This edge-to-edge plowing, together with fertilizers, insecticides, and herbicides, left scant space for wild plants or animals. Nature had been displaced in major crop areas as completely as it had been displaced in urban centers. This was especially true for prime agricultural soils, whether they were used for annual or perennial crops, whether for grain, cabbages, or grapes.

This transformation did not apply to all rural land. Permanent pasture did not change much and still supported a variety of plants and wildlife. Most of our forests were not intensively managed, although forestry in the most productive areas in the Southeast and Northwest was becoming more like modern cropland agriculture.

The loss of the best soils to urbanization was America's greatest concern. Under modern methods of farming, however, some of their natural values were already lost. Conversion to development added only the feature of making the loss permanent.

If intensely cropped land were converted to low-density suburban development, the natural values of the land could be improved. The diverse landscaping, the decrease in use of herbicides, and the feeding of wildlife could lead to the support of many more kinds of plants and animals than were there before.

In Britain, land was farmed ever more intensely throughout the 1970s using more fertilizers, pesticides, and herbicides. Fields grew larger and hedgerows disappeared at the rate of 200 miles of hedge a year. Suburban development, in contrast, placed more emphasis on vegetation, both natural and horticultural. Some ornithologists considered at that time that the future major habitat of many British songbirds and other wildlife would be the suburbs, not the farmlands.

AGRICULTURE AND LAND

Although the major concern was with urbanization, agriculture transformed much more land than did development. The 465 million acres (more or less) of land used for crops in 1980 came mostly from forest, range, and wetland. This removed about 25 percent of the original forest (more was removed but has since grown back), 30 percent of the original rangeland, and 40 percent of the original wetland. Furthermore, the best soils were converted, reducing the average productivity of the remaining forest and range.

In addition, about 5 percent of the original forest was converted to permanent pasture, and much of the remaining rangeland was used for pasture.

Land was still being converted to agriculture in the late 1970s. The two major processes were by drainage and irrigation. The Department of Agriculture estimated in 1958 that about 107 million acres of poorly drained land (including wetland) had been drained and that about 44 million acres of potentially good cropland remained to be drained. These drained soils were highly organic and usually excellent for intensive agriculture. Drainage continued, in the Southeast, Midwest, Northern Plains, and in California, but there are only poor data on the annual increments that resulted.

In 1974 about forty-one million acres of farmland were irrigated, an increase of twenty million since 1940. Most of this was cropland which used to be rangeland with inadequate or unreliable rainfall.

These additions of cropland may have balanced losses to urbanization, since USDA estimates show that the supply of cropland changed little in the twentieth century. Urbanization of cropland could, in the future, increase the demand for draining wetlands and for expanding irrigation.

A REGIONAL DIVERSITY OF LAND

People prefer variety in their landscapes, mixtures of forest, pasture, cropland, water, wetland, hills, and streams. Even the pleasure of hiking deep into a large forest depends on the contrast with somewhere else; the early settlers pushing westward through the forest learned to hate its damp and gloom.

Diverse landscapes also support more kinds of plants and animals. Edges between different kinds of vegetation, such as between field and forest, are especially supportive of wildlife. Edges not only share the species found on both sides, but are the habitat for species that require both kinds.

For these reasons, the mixtures of land types present in a region are an important environmental concern. Whatever diversity is present should be conserved. Indeed, sometimes it can be increased. Where urbanization is involved, natural values are served better by using land types that are regionally common rather than those that are regionally scarce.

The composition of land varies greatly across the United States. Ten areas, listed in Table 1, represent both the general geography and the extremes of conditions that were found in the early 1970s. For each state, the rural land not already committed to a special use (highways, parks, defense land, etc.) is considered. It consists of cropland, rangeland, forest, and "other" land, shown as a percentage composition in Table 1. "Other" land is mostly wetland, except in Arizona and California where part of it is barren desert or mountaintop. About a fourth of this land is state or federally owned (but not reserved for a special use). The rest is privately owned. Presumably any of it could be urbanized if conditions were appropriate.

Also shown in Table 1 is the percentage of this same land that is in farms, varying from 13 percent in New England to 96 percent in Iowa.

In New England, pasture and cropland are both very scarce. These should be preserved against conversion to any other use. Forest exists in abundance. Not only should forest be used for urbanization, but the diversity of land would be improved if substantial areas of forest were put into pasture (as they were not too long ago).

TABLE 1. THE COMPOSITION OF RURAL LAND

Region	Cropland	Rangeland	Forest	Other	Percent in Farms
New England	6	1	87	6	13
Georgia	21	5	72	2	40
Ohio	56	7	27	10	66
Mississippi	31	9	58	2	49
Iowa	83	6	7	4	96
Texas	24	59	15	2	83
Wyoming	5	81	10	4	60
Arizona	3	64	27	6	59
Washington	22	18	55	5	45
California	13	28	47	12	40
48 States	26	34	34	6	58

Source: USDA Agricultural Economics Report, No. 440, 1974.

Note: Rural land already committed to a special use is excluded. The remaining land is broken into its percentage composition. Other land is mostly wetland, includes barren desert and mountaintops.

In Iowa, pasture and forest are scarce. Using these for urbanization would worsen an already poor situation. But most of the cropland is of very high quality and should be preserved as an agricultural resource. Clearly, Iowa does not have any "spare" land. The best solution to the problem of urbanization is not to do it.

Wyoming is the third extreme example. Cropland and forest are scarce. Much of the range is publicly owned. Wyoming, with its low population density and few demands for urbanization, has an abundance of open land that could be urbanized.

States like Texas, Washington, and California are diverse with respect to cropland, range, and forest. They have, therefore, more freedom in the land they choose for urbanization without jeopardizing this particular aspect of environmental value.

Because about 60 percent of the land available for urbanization was farmland, often near urban centers, it contributed the majority of land used for past urbanization, and the same pattern seems to be forecast for the future. Of this farmland, 45 percent was in cropland, 36 percent pasture and range, 16 percent forest, 2 percent wetland, and 1 percent farmsteads in the late 1970s. Another 16 percent was nonfarm, privately owned land. It was mostly forest, with wetlands included, and also contributed land for urbanization. The remaining 24 percent of available land was state or federally owned. It was roughly equal amounts of range and forest, plus wetlands. Although much of it was remote from urban areas, some was used for development each year.

ENDANGERED ECOSYSTEMS

The land vegetations of the forty-eight conterminous states are classified into 106 natural types or ecosystems. Various portions of these are left in various locations. Twenty-three ecosystems have lost more than a fourth of their original areas; thirteen, more than half; and ten, more than three-fourths.

Elm-ash forests (and their various successional stages) once covered 5.5 million acres in Ohio, Indiana, and Michigan; 12 percent remained in the 1970s. The sand-pine scrub of Florida once covered more than 60,000 acres; 15 percent was left. Of the original 350,000 acres of tule marsh in California, Utah, and Nevada, all but 11 percent was destroyed. Conversion to agriculture was the major cause for these past losses, but the surviving fragments of all three are threatened by both agriculture and urbanization. In California, urbanization was the major destroyer of at least three ecosystems: California steppe (69 percent gone), mesquite-live oak savannah (39 percent gone), and coastal sagebrush (36 percent gone). And urbanization still spreads.

Eleven of the forty-eight states still have over 80 percent of their natural vegetation. Eight of these are the mountain states. At the other extreme, the remaining natural vegetation retained in Indiana is 18 percent; in Illinois, 11 percent; in Iowa, 8 percent; and in the District of Columbia, 0 percent.

Wetlands suffered similarly. Of the original 127 million acres, about 70 million remained in the 1970s. This was less than 4 percent of the land area, although it varied from 2 to 10 percent, as shown in Table 1. Wetlands are the dominant support of waterfowl, a variety of mammals, and many unusual plants. Some wetlands are among the most productive ecosystems known. Wetlands depollute water and reduce flooding. In 1972 the Army Corps of Engineers found that the natural wetlands of the upper portion of the Charles River in Massachusetts were a better protection against floods than the construction of any likely alternative using detention reservoirs or dykes.

Drainage of wetlands for agriculture has been noted. Coastal wetlands were destroyed by urbanization and other forms of coastal development, especially along the eastern megalopolis and in the San Francisco Bay area. Until the widespread wetlands legislation of the late 1970s, almost all wetlands and ponds caught up in urbanization were filled and built over.

By 1980 many states were trying to protect wetlands, and the nation was mounting an effort to strengthen such programs. This could affect the direction in which urbanization spread, not only to avoid wetlands but to avoid mixed landscapes of wetland and development. The close proximity of neither benefits the other; people do not like mosquitoes, and wetlands suffer from close development.

The whole subject of endangered ecosystems was one of national concern in the late 1970s, and an effort is underway to ensure that some portion of each kind of ecosystem is preserved. Although the national impact on land use would be small, the conservation of ecosystems could have significant local effects, especially in the Corn Belt, where so much land makes such good cropland and where relatively high population densities make other demands on land.

The nation and many of the states passed legislation protecting the habitats of species that were determined to be threatened by extinction. The snail darter and furbish lousewort are famous examples. Most rare or endangered species occur only in a few places, in very particular habitats, although a few, like the bald eagle, are widely distributed. When the only home of a species fell within an urbanizing region, conflicts arose.

Usually the protection of these species near urban centers was included in larger conservation programs and was used as an added argument in favor of such programs. The San Francisco Bay National Wildlife Refuge is one such plan. It began in the early 1970s the process of protecting 23,000 acres of estuarine wildlife habitat in the bay region after two-thirds of the original area had been destroyed. The proposed refuge included habitats of four major rare and endangered species, three kinds of birds, and one species of mouse.

Cases involving a single species are not numerous. A spectacular white salamander was found within the urbanizing region of San Marcos, Texas. It inhabited underground streams associated with Purgatory Creek. The Nature Conservancy purchased one cave site, but the issue of protecting water quality throughout the habitat remained. Conservationists and developers were repeatedly at odds with respect to one land parcel after another.

Another example unfolded in the winter of 1979–80 in Plymouth, Massachusetts. Although Plymouth was the first settlement in the northeastern United States, it remained mostly rural until 1950. Then it was reached by the expanding periphery of greater Boston, and development pressures accelerated. One large area of deep sandy soil supporting pine forests was especially "developable." This region also included several ponds that were the only known habitats of the Plymouth red-bellied turtle, which had southern cousins but was the only large basking turtle in New England. In 1978 the Nature Conservancy acquired 180 acres of this land, including some ponds. In 1980 the U.S. Fish and Wildlife Service proposed that 7,000 acres be declared "critical habitat." If this were implemented, further development would be rigorously controlled, creating further conflicts between conservationists and developers.

One effect of restricting development in some locations in order to protect rare and endangered species would be to accelerate development elsewhere. "Elsewhere" could be cropland or farmland, since endangered species are usually restricted to undisturbed natural habitats, i.e., on wild lands. While this could be a severe local problem, the overall quantitative effect would be small.

Land for Recreation

Some land for recreation was created within urban land as it spread across rural areas. In most land use surveys, urban land includes playing fields, parks, golf courses, and natural areas.

AT THE URBAN FRINGE

Land for recreation adds to land used for other activities and increases the rate at which land is urbanized. Such open spaces are often located early while large parcels are available and before land prices are too high.

Golf courses are a significant fraction of urban recreation land. They generally require flat or moderately rolling land with soils suitable for lawns. Open land is preferred, since deforestation and stump removal are usually too expensive in the economics of golfing. Thus, golf courses are very likely to replace good farmland. In the 1960s two new golf courses were built in a rapidly urbanizing region southeast of Boston. One was created out of pasture, old fields, and a little cropland, before development had reached that area. The other replaced an aggregation of cranberry bogs that had become surrounded by development. Although cranberry farming was the most profitable form of agriculture in this region, it did not survive the pressures of urbanization. In New England, cranberry bogs are not bogs. They were built by covering low land with sandy fill, creating flat surfaces and a system of ditches and small reservoirs. The bogs are seasonally flooded and drained to promote maximum cranberry production. Changes needed to convert cranberry land to a golf course are inexpensive. Because muck or peat usually is beneath the fill, bogs offer poor support for major construction projects. In this urbanizing region, golf courses account for 2 percent of all urbanized land.

Land for parks and natural areas is classed as recreation land if the surrounds are rural and is classified as urban if the surrounds are urban. Thus, urban growth includes to some degree the reclassification of such land without a change in land use. In addition, however, new land is designated at the urban fringe for recreational use.

The kinds of land preferred are variable. Natural areas include water, wetlands, and bluffs or outcroppings—land not well suited to more intensive development. They also include woodlands or other good local examples of natural vegetation. Such areas would, without protection, eventually be developed; in the last stages of infill development, any land not protected becomes developed. Reserving natural areas pushes urbanization outward.

Parks also often include land not well suited to development. But a large proportion of urban and suburban parkland is flat and open, the same kind of land that could be used for golf courses. Parks are, in fact, an integral part of urbanization. They make denser residential development more acceptable and, in that sense, may not increase the overall land requirements for urbanization. In any event, few would seriously recommend omitting land for parks in order to reduce the amount of land being urbanized.

A general planning goal in the 1970s was to provide seven acres of urban park for each 1,000 people. Most American urban centers fell short of this. Even so, for an urban population of about 150 million people, 1.05 million acres of park would meet the goal. This was about 3 percent of the total area of the United States urbanized by the late 1970s.

GUIDING THE FRINGE

Although land for urban recreation amounts only to about 4 or 5 percent of urban land and has a small effect on the amount of land urbanized, it could have a considerable effect on the patterns of regional development. When land at the edge of development is designated as park, all land adjacent to the park becomes more attractive. People prefer to live next to public open space.

The state of Michigan began some years ago acquiring recreation land all over the state, especially in regions expecting urbanization.

Near greater Detroit the state acquired land for a corridor park along the Huron River. This action accelerated parceling and development of land adjacent to the corridor. The frontages were developed first and sold at high prices, leading to fingers of upper-income residential housing pushing into rural areas. The rest of urbanization followed. Thus, park programs could be initiated in expectation that urbanization would follow. This could direct development away from prime agricultural land.

REGIONAL RESOURCES

Some kinds of recreation land cannot be included within urban areas or at the urban fringe. Areas with primitive camping facilities are not appropriate near cities where they would be subject to casual use. Outstanding natural areas occur where nature put them. They are much harder to preserve near cities. Both wilderness and "natural" areas are needed to complete the recreational opportunities demanded by urban people.

At the national level, a variety of national lands serve these purposes. The amount of such land has been growing more rapidly than urban land. Creating these open spaces had little impact on either urbanization or agriculture. They tended to be created where the land had little other value, especially in the mountain states, in mountainous areas of the Pacific states, and in scattered other mountainous and seashore areas. Very little cropland was involved, and both the rangeland and the forests were usually of low value.

Such lands are also needed at the regional level, primarily as state parks and reserves. The existence of such lands, and the opportunity for creating more, varies greatly from one region to another. Ten regions comprising the forty-eight contiguous states are shown in Table 2. For each region, the land available is expressed per 100 people in the region, as a means of expressing the amount of land in relation to the regional demand.

Parks were identified by the 1970 Census as national and state parks, forest preserves, and primitive areas. Wildlife refuges were identified as areas administered by state and federal wildlife agencies. "Water" included all inland water, excluding the oceans and the Great Lakes. The total water area increased as reservoirs were created, perhaps at the rate of 200,000 acres per year in the 1970s.

TABLE 2. ACRES OF LAND PER 100 PEOPLE LIVING IN EACH REGION

Region	Parks	Wildlife Refuges	Inland Water	Forest	Range
Northeast	7	3	8	129	6
Appalachia	7	4	22	393	40
Southeast	12	6	26	438	65
Lake States	11	10	29	306	31
Corn Belt	1	2	4	84	40
Delta States	2	12	43	648	96
Northern Plains	10	17	47	90	1,461
Southern Plains	9	5	30	242	815
Mountain States	247	49	61	1,447	3,711
Pacific States	45	4	12	353	211
48 States	22	7	18	296	294

Source: USDA Agricultural Economic Report, No. 440, 1974.

Northeast: ME, NH, VT, MA, RI, CT, NY, NJ, PA, DE, MD, DC. 116,514 thousand acres.

Appalachia: VA, WV, NC, KY, TE. 128,148 thousand acres.

Southeast: SC, GA, FL, AL. 128,064 thousand acres.

Lake States: MI, WI, MN. 127,000 thousand acres.

Corn Belt: OH, IN, IL, IA, MO. 166,329 thousand acres.

Delta States: MS, AR, LA. 95,579 thousand acres.

Northern Plains: ND, SD, NB, KA. 196,614 thousand acres.

Southern Plains: OK, TX. 215,844 thousand acres.

Mountain States: MT, ID, WY, CO, NM, AR, UT, NE. 552,886 thousand acres.

Pacific States: WA, OR, CA. 207,274 thousand acres.

The Corn Belt seems outstandingly deficient, due to the combination of a high population and a lot of cropland. Creating a sufficient area of recreation land would put serious pressure on the remaining forests and pastures. Should cropland be converted to recreation land? Or should people accept the necessity of traveling further to find recreational resources in other regions?

The Northeast has large acreages of parks and water, but a very large population, indicating a level of congestion that is all too familiar to the residents of the region. The delta states have lots of water but little park, while the Pacific states have lots of park and little inland water.

All of these numbers, based on the 1970 census, have shrunk, on the average, about 10 percent during the last decade due to population growth. Movements of people during the last decade shrank numbers more in some regions than in others.

The Carbon Dioxide Problem

The amount of carbon dioxide in the atmosphere increased in the 1970s. People were almost certainly the cause, since the carbon dioxide produced by burning fuel was larger than the amount added to the atmosphere; some was probably absorbed by the oceans. Climate theorists generally agree that higher levels of carbon dioxide warm up the planet, changing climates everywhere. They also generally agree that the level of carbon dioxide would decline very slowly if America and other countries stopped burning fossil fuel.

Burning fuel is not the only way that people produce large amounts of carbon dioxide. Expanding population has reduced the world area of forest. The draining and plowing of highly organic muck soils increases their rates of decomposition ("burning"). And in all highly organic soils the use of fertilizers increases the "burning" rate. By these and other activities, substantial additions were made to the carbon dioxide produced from burning fuel. Thus, humans may cause climate changes which could last a long time if the present trend continues.

What role did America play in this problem? The United States during the 1970s burned 27 percent of the fuel consumed each year. The U.S. contributed little or nothing by deforestation; it was balanced by reforestation. America drained over 100 million acres of land and each year fertilized these and most other highly organic soils that used to support tall-grass prairie. This caused an undetermined addition of carbon dioxide to the air.

The problem is serious and will be of long duration. But it has little to do with urbanization in America. If more forestland is urbanized, carbon dioxide will be released as the forest is destroyed. If more cropland is converted to suburbs, the amount of vegetation used in landscaping would probably be larger than it was before under crops, which would remove some carbon dioxide from the air. But the amounts involved would be trivial. America, at the 1979 rate, was burning coal, oil, and gas at a rate equivalent to burning up twenty million acres of mature forest a year.

Conclusions

Converting cropland into cities has little direct effect on natural values. By 1980 neither cropland nor cities had much natural value left.

Additional uses of land at the urban fringe for recreation, conservation, and species protection are each small but, taken collectively, make a moderate increase in the rate at which fringe land is converted to a special use.

The composition of land available for urbanization varies greatly from region to region. Problems are most severe where large populations require a lot of urban land or where most of the land is good cropland. These two conditions converge in the Corn Belt.

Losses of cropland near cities increase pressures for draining wetlands elsewhere. Agriculture has already drained something like 40 percent of the nation's wetlands.

The regional amount and composition of land available for rural recreation, compared with the sizes of regional populations, varies greatly. Opportunities for meeting new demands are low where population levels are high, reducing the total amount of land per 100 persons. The problem is worse where high populations exist in regions of abundant good cropland.

C. Lowell Harriss

7

Free Market Allocation
of Land Resources:

(What the Free Market Can and Cannot Do in Land Policy)

In the long run, in spite of a much-quoted quip of Lord Keynes, we are *not* all dead. In any realistic sense, in any humane and human meaning of life, "we" constitute a continuing group. Human beings have always tried to prepare for a future which would extend indefinitely beyond their lives, and all civilization has rested on actions which assumed a future. These actions have not always been the best possible. Avoidable losses have resulted from short-sighted failures to prepare for what might reasonably have been foreseen.

Introduction

Land allocation presents problems to human beings, here and now, today and in the future, and no "solution" emerges that is

C. LOWELL HARRISS *is professor of economics at Columbia University and economic consultant at the Tax Foundation, Inc. In 1972–73 he was president of the National Tax Association-Tax Institute of America, and from 1972–78 he was vice president of the International Institute of Public Finance. Dr. Harriss has been a consultant to government, a member of several public and private economic associations, and has written many books and articles chiefly on aspects of public finance. He also edited the American Assembly volume* The Good Earth of America: Planning Our Land Use.

demonstrably correct; nor do these problems lend themselves to decisions which are precisely "right or wrong." Some combinations of results, judged by widely accepted standards, are better than others. Issues involve not only participants directly concerned but others as well. Most "solutions" are compromises that offer the best prospects of getting farthest toward the desired end of a spectrum that extends from horrible to ideal. Some "solutions" are subject to much criticism.

Everything that has been done (except sun, rain, wind, and other works of nature) is the result of human decisions. People act as individuals and as groups. Among the groups are governments—local, state, and national. One concern of this chapter is to consider if and when governmental decisions can be expected to produce better results than private decisions made in the free market.

The concept of a "free market" for real estate is less valid than for commodities, shares of common stock, or human services. American government, chiefly state and local, has, since about 1900, imposed more and more restrictions on the private use of land. The collective processes of government influence actions in real estate more than most market transactions. Nevertheless, terms of purchase and sale or lease of particular parcels remain open for bargaining.

Special Characteristics of the Land Market

LOCATION

The real estate market, like other markets, involves individuals and voluntary groupings (corporations, for example) making and receiving offers. But land differs from all other commodities. Each unit actually or potentially exchanged differs from all others in at least one respect—location. Thus the neighborhood has significance without counterpart in most transactions. Units are not portable. Forces making worth (or value) or detracting from it can transfer economic desirability. Chapter 1 points to some of these forces.

The *local* nature of realty affects the market. Houses may be in demand in one city or one part of a city and in surplus in another. For decades, sources outside the community have financed many real estate transactions. The development of a national and even an international market for lending, purchase, development, and operation has influenced decisions that might once have been purely local, but these nonlocal forces could not escape local limitations. A nice building site in the "wrong" town could not be moved to the "right" one. Buildings could be moved only if they were quite small and the distances quite short. Even under the best conditions the cost was very high.

Improved communications since about the time of World War I extended the geographical scope of all important real estate markets. This improved the effectiveness of the market, but distant participants seldom had local voting power and showed little concern for local considerations. They had broader visions, greater ambitions, and more adequate financing than most local people. Outsiders' bargaining power came from the eagerness with which communities competed for employers and tax base. Decisions to build a new facility, a factory for example, were made in far-off counting houses, but such a decision could fundamentally change a town for generations. A decision to abandon a facility, leaving it empty and unsalable, could injure a town for years. (The decision of Amoskeag Corporation to leave Manchester, New Hampshire, was such an instance.)

The *lack of homogeneity* of product distinguishes this market from any other. The units are often large relative to those of most markets. *Financing* is a problem and, as Chapter 1 indicates, mortgage practices influence real estate development. Leasing can serve instead of sale, and either way, payment becomes a series (a flow) rather than a capital sum (a stock). Government regulation can alter values without much warning.

In summary, this market has special features which prevent transferring to it many of the elements properly applied to the "free market."

FUTURE UNCERTAINTIES AND FIXITY OF SUPPLY

Real estate embodies two of the most difficult elements of economics in the "real world" sense—space and time.

Future human desires can be adumbrated but cannot be fully reflected in "today's" market, whenever "today" is, so that information relevant in the long run is inescapably incomplete. Future voters cannot express preferences "now." The available knowledge about the future cannot guide market decisions about land as it can about nondurable items like clothing which may not last until next year. Land lasts forever, and buildings last for decades. The human beings who will participate *then* cannot express their views *now*, and those who make decisions *now* must remain uninformed about many aspects of their own circumstances later in life.

A free market operating as well as can be expected cannot possibly embody judgments which require balancing of facts that are yet to develop, but many economic decisions depend on judgments about the future. Men and women "today" do try to recognize uncertainties and take unknowns into account. Family and corporate saving and investment look to the future even though the investors do not see it fully. Community actions also include preparation for years ahead, some explicit and some implicit.

Many forward-looking decisions, family, business, and governmental, deal directly with land use. Every event of life involves some use of space on the earth's surface, but future-oriented actions have generally given little thought to what will happen to land. They assume that others will take care of the specific details.

Specialization, which contributes enormously to the high standard of American living, restricts each person to only a tiny fraction of those things that contribute to a family's living. People generally rely on others for what they get now and, for most, planning for what they shall want later. They exchange, as do groups, including local governments. This involves cooperation in myriad ways and frees people from the need to manage most inputs to their consumption. Consumers select from among the alternatives which others offer. Decisions about land in the future may be one case in which "others" have wide latitude to act because individuals cannot do everything for themselves. Clearly, today's land markets cannot embody the preferences of a future extending as long as land will be needed.

FIXITY AND INVARIABILITY OF SUPPLY

The "space" characteristic of land involves more than location and does so in ways that further distinguish land from other productive resources. This fixity of quantity has special significance for actions anticipating the future. Although for any one use, the quantity of land is adjustable, the total is not substantially expansible. Major topographic features like mountains, oceans, and deserts have seldom been changed by market forces. The fixity requires deliberate attention.

The total of space on the earth's surface is subject to little change.

The same normally applies to the total within any political entity, nation, state, or town. The supply of labor expands, capital goods are produced, and the quantity of man-made products increases as more units are added, both new types and familiar ones. Land does not expand, even though the price paid per square foot may rise more or less continuously, and at times by large amounts. Some "nonuseful" land can be made useful by filling swamps or ponds or cutting down hills, but the investments of funds and effort required represent man-made capital.

Effects of Price Change—Prices affect the amount available for any one kind of use as against others. Within its overall constraint, supply is variable for each of many kinds of uses. The market allocates land by price. Each participant in the market subject to government restrictions has a chance to pay the market price and can get space for some specific purpose. Each particular user faces this now and will expect to in the future.

Individual farmers and businesses are concerned with particular parcels for particular purposes and have little reason to worry about the fixity of total land supply. Their concern is about land for specific uses. If one parcel is not available, another will usually (but not always) serve as well.

Bidders ordinarily have several alternatives. Choices may be limited and particular needs hard to fill, especially where zoning or other governmental rules outlaw uses or block transfer from one use to another. Private business, lacking the power of eminent domain, may not be able to assemble a large plot "in town" at any price, or at a price which is economically acceptable.

General Effectiveness of the Market in Allocation—Buyers and sellers of land thought and acted in terms of potential expansion for *particular* uses and had to make adjustments as conditions changed. The desire to maximize profit had tended to move land into the use which would be most profitable for the foreseeable future. In this way, individual actors in the market look to the future. Yet "present" society, as part of a continuing humanity, has an obligation to take account of the basic reality of land limitation. Observers classifying areas suitable for various uses, such as agriculture, reach conclusions about the area actually or potentially available; but published figures may be questioned for reasons noted by other authors in this volume. Classification has presented problems. Flexibility among uses might be greater than recognized because price changes make a difference, and technology and capital formation exert influences. But the use of a specific parcel for one purpose always prevents its use for any other. And in contrast with capital goods, the price paid for land does not finance the creation of new land.

The free market forces the buyer of a plot of land to pay as much as it would be worth for the best alternative use. Real estate markets have many imperfections, especially when compared with the ideal of economic theory; but allocation reflects the best judgments of those close to the facts. Directions of change are determined by participants acting at the margin where relatively small changes exert significant influence. Emotion, hopes and fears, speculative fever, and financing opportunities all operate, and available information and desire to use funds to best advantage affect choices made and choices passed by. The "market" should not be romanticized as an approximation of perfection, but it does allow human participants to test, compete, and choose.

Nevertheless, a major problem grew out of that peculiar characteristic of land: "they are not making more of it." Actors in the future are not present now to express their views about alternatives then regarding the opportunity costs of the future. The grandchildren of any generation cannot meet their wishes for land by creating more of it. Successors could build factories and electric generating plants as conditions indicated, but not land. As the years passed and new conditions emerged, descendants, the present generation for example, had to get along with the land they received, and the same will apply after the present generation is long since gone.

The essential nonexpansibility of land has been recognized for decades. Astute investors of the past took it into account. But serious questions emerged in the late 1970s. Did the generations of the 1870s and 1920s see as clearly as conditions would reasonably permit? Can present society through government improve on the results which the free market would otherwise produce in guiding land use?

IRREVERSIBILITY: A FACTOR ALTERING PRIVATE MARKET EFFECTIVENESS

Once land has been taken from one use, returning to that use later has been exceptionally difficult. This characteristic goes along with fixity of supply. As Chapter 1 notes, shift from one crop to another might be easy, but shift from highway or shopping mall to food production is entirely different.

Typically, the kind of resources employed to put land to use creates obstacles to reversion to earlier use. More specifically, housing, commercial buildings, airports, factories, schools, and other changes, as land is shifted out of agriculture, require commitments that preclude a return to farming.

Admittedly, transfer back *could* be made at some cost. Abandoned farms might be cultivated again. Even Times Square could be returned to farming. New towns or suburbs could be reconverted to vegetable fields or dairy land. The cost depends upon the complex of factors of each case. Often the expense is utterly beyond anything economically sustainable. The shift of land use from farming has almost always involved combination with structure, and the capital added was essentially immobile. The new investment could be very large relative to the value of the land and could change the fundamental nature of the property. To a large extent, the transfer of land away from agriculture has been irreversible.

An acre in exurbia worth around $2,000 for farming in 1980 might be supplied with roads, sewers, water, and "improved" with four houses each selling for $75,000. The package would then be worth $300,000. Even if farm values increase ten times over, the cost of returning the land to farming could not be met, and it would be a lot more than the cost of keeping what developers call "raw land" in farming in the first place. The addition of capital which is not movable fundamentally changes the property. What was an asset of modest amount, farmland, becomes an integral part of a much more valuable property as capital is added. The resources which are combined with the land might be used in any of several places. When they are applied to a particular parcel, the fluidity ends, and long-term fixity sets in.

Speculative Holding for New Use—The free market includes speculation. This activity has often been justifiably maligned, but it has served purposes useful for the economy. The speculative market gave a farmer a chance to sell between the time farming became uneconomical and the time a developer wanted the land. Few developers stockpiled land; most bought land only as they needed it. Hence, there were long "ripening periods" between the end of economically sensible farming and the time of actual development.

Speculators have been amply motivated to keep land in uses below the short-run optimum in the belief that the future would present better opportunities. The conditions are not those of commodities and other futures markets except for possible efficient short-run speculation. The gap between "present" and "future" land markets is one of years, even decades. Continuity in this sense exists in few commodities. In land trading, if an estimate of the worth of alternatives turns out to be wrong and the future brings conditions other than those expected, the error might be recognized but could not be corrected because of the irreversibility factor. Hedging and arbitrage are possible in other futures markets, but have rarely been possible in the real estate market.

Financial markets are generally rational and efficient, in that knowledge is rapidly embodied in securities prices, although expectations about the future could differ even when fully adjusted to existing knowledge. Similar forces also operate in real estate markets, but conditions have so many special elements, compared with securities markets, that land prices at any time include more discrepancies, i.e., failures to utilize knowledge fully.

As new conditions develop, alert real estate operators always try to take advantage of the latest information, but their ability to do so is limited. The adjustment is quite different from the nearly instantaneous sale of one security and the purchase of another in an organized financial market. Each piece of real estate is unique, if only for its specific location. Markets are largely local, and prospects of zoning create uncertainties. Financing is sometimes complex and, in any case, is specific to each parcel. Income tax factors differ from one person to another. Most users of the future, even the near future, cannot participate in prior speculation. Free market speculation might have some merits, but it seldom takes place in conditions which recognize future interests as fully as society could realistically desire.

DISCOUNT RATE

Decisions about uses in the future require, implicitly or explicitly, a discount rate to allow for the cost of using resources through time. Markets regularly have produced interest, and discount rates have reflected the judgments of all participants at the time, active and potential. Rates, especially long-term rates, have traditionally been fairly stable, but in times of uncertainty, compounded by inflation, rates have shifted significantly from year to year and from month to month.

Periodic differences have substantially influenced long-term projects. The present value of a dollar due in twenty years is about thirty-two cents at a 6 percent interest rate but just over ten cents at a 12 percent rate. The constant monthly payment required to amortize a $100,000 loan in twenty years at 6 percent is about $716. At 12 percent, the same payment is over $1,100. The 12 percent payment is less than twice the 6 percent payment because of the relationship between declining interest payments and increasing amortization, the complex details of which would be an inappropriate intrusion in this commentary. Interest rates fixed in former markets grossly affect prices later on. Projects financed before 1975 often carried long-term rates of 6 percent or less. In the 12 percent market of 1980 these 6 percenters sold at a premium. The seller was "selling" the mortgage as well as the property, and if supplementary financing was required, a "wrap around" could usually be negotiated, preserving as much of the low rate as possible.

Many mortgage lenders rely on "prepayment clauses" to refuse repayment of high rate loans when rates fall. These clauses keep a borrower from repaying more than a fraction, often 20 percent, in any one year. Property with high-rate loans and restrictive prepayment clauses sell at a discount compared with projects financed later at lower rates. Judgment on interest rates casts its shadow over a long future.

There is little evidence that collective decisions by government are better than the market. They would be better only if public officials acted more prudently in their official capacities than they did in their private affairs, or if persons too weak to influence the market as individuals could do better than their own ability or willingness to put their own funds on the line.

Private market decisions are made by persons committing their own funds. In government decisions the connection between personal commitment and expected result is more remote. Emotion, ignorance, venality, and the abuse of power combine with many other factors to influence political and bureaucratic actions. Government actions have usually (but not always) involved good intentions, consideration for the financially weak, and sometimes great visions of future benefits, but they have often grossly underestimated costs. Political decision makers have sometimes taken a longer view than private operators, reflected a greater sense of history, and shown more respect for the future. Rates for government projects have sometimes appropriately been under market rates, especially where some irreversible action was involved, like winning a war or maintaining a food supply. If the market valued the future on the basis of 10 percent while government agencies would lend at 7 percent, then the government could promote and finance projects which the market would reject.

By making a special rate available only in certain places or on certain types of projects, government influenced the whole construction business in the 1930s, 1940s, and 1950s. The hidden cost was easily overlooked, if only because dollar figures were not attached.

Many federal government projects involving off-market discount rates left much to be desired. Wasteful undertakings were pursued with distressing frequency—distressing at least in the absence of reliable justification for apparently unwise sacrifices of the present for the future or of the future for the present.

In summary, the market generally has done a better job in setting discount rates than politics has, but the best the market will do cannot adequately reflect valuations of the future for a commodity like land where supply is fixed and uses are not reversible at tolerable expense.

ASSEMBLY OF PLOTS: EMINENT DOMAIN

The nature of land and its use gives rise to another significant factor peculiar to the private versus government issue. Highest and best uses often require plot sizes different from those now existing. Private owners can divide large units into small ones within the scope of zoning ordinances and deed restrictions. Existing structures have often presented problems, especially those immediately adjacent, because an owner owes to his neighbors the "right of support." He cannot, for example, dig a foundation for a new building in such a way that a neighboring building falls into the hole. But private market forces could almost always change a large plot into smaller ones.

Private operators have had far greater problems in assembling small plots into large ones. "Holdouts," whatever the reasons, greatly raised the costs. Assemblage was a well developed art in the early part of the century, but after World War II, better highways enabled operators needing large plots to save money by going to exurbia and buying farms.

The owner of land needed for combination with other plots could and often did prevent a project entirely. One or a few owners often made it more expensive than inherent realities required (vague as the concept of "inherent requirement" may be). The owner of one rather small unit could impose costs that altered the nature of the final result. This was a sort of market failure created by locational conditions imposed on the private market in land. The many different checkerboard and leapfrogging patterns around cities materially affected the potential for efficient food production.

In contrast, government has the power of eminent domain through which it can compel results. Government retains this potential authority which is not matched by bidders in the private market, although it is occasionally given to a regulated utility like an electric company which could not acquire a right of way without it. The practical significance of eminent domain for the future depends on a variety of factors, including the foresight with which it is used.

The importance of location can exact economically unjustified costs which have relatively great significance for specific projects. Government can use eminent domain wisely to achieve results far superior to those of a market in which holdouts exert exploitative power. For the issue of farming versus nonfarming, a few holdout parcels can affect the most productive use of large areas. The development of fractured areas involves costs far above the costs of developing compact areas. Farms left over in fractured areas are harder to operate than others in areas of uninterrupted farming.

MOBILIZING FUNDS

Some land development projects involve large sums of money. Private markets sometimes mobilize funds on an adequate scale for even an ambitious undertaking. In other cases, private sources cannot raise enough money to do something that many persons consider desirable. Similarly, when governments are involved, either condition can exist. In some cases, government, national, state, or local, can borrow and lend or can insure private loans.

Government, chiefly federal, has been involved in real estate financing at least since the Federal Land Banks were established by legislation early in the Woodrow Wilson administration. Resources raised by sale of bonds were loaned to farmers. These banks were modeled on the German Land Bank system. The hearings, studies, and preparation were accomplished under Dwight Morrow, a one-time Morgan partner and a U.S. senator while Taft was President. Prior to the establishment of the banks, interest rates on farms were very high, and credit was unobtainable in secondary locations. Thereafter, rates were equalized geographically, and funds were available in all good farm areas. On the less desirable side, available credit helped propel the rise in farmland prices that plagued the country from about 1928 to the mid-1930s. The loans were capitalized into values, and, to that extent, the market for land was less "freely competitive" in the strict sense.

Agricultural price supports have been in existence since about 1935, and the total aid further raised the prices of farmland. Not all kinds of farming were equally affected. Many other forces acted on farmland prices. Federal intervention in the mortgage market for housing after 1935 was one of the great forces decentralizing cities and obviously influenced land prices on the urban fringe, land which was teetering between farm and nonagricultural uses. Conditions required for a fully free market did not exist for almost seventy years in American agricultural communities.

Pointing out that this market fell short of an "approximate ideal" free market does not automatically imply that more governmental action would have improved results. The power to do better has existed. Using it wisely required politicians and bureaucrats to act more wisely than they did.

GOVERNMENT'S USE OF COMPULSION AFFECTING LAND MARKETS

Government has power which operators in the private market have not. Government uses compulsion in the form of legal restraints and zoning and through credit vehicles like the Federal Land Banks, the Federal Housing Administration, and other mechanisms which reflect no small amount of human ingenuity. Such authority can alter land use without requiring a cash outlay. A lack of symmetry certainly exists between government and the market.

The use of powers of government, including putting cash on the barrelhead, has influenced current payment for future benefits often without necessarily compensating owners adversely affected. Private parties also have done this but not to the same extent. Government has long influenced land use, favoring some owners and hurting others, usually intending to benefit some group deemed underprivileged; but a benefit conferred in one place has usually been matched by a deprivation elsewhere. Much can be done without compensating private owners, an example being the construction of a limited access superhighway parallel to a once busy commercial stretch of highway. In such cases and to varying degrees, they exist almost everywhere. The market has not operated freely. Yet in adjusting to changes in regulation, the market has always taken account of the available elements of freedom.

Inducements to delay the conversion of farmland to other uses offer an example. Participants in the private market would not act as they did if government had not set the rules it did. The effects depend on local conditions and, obviously, vary from place to place.

EXTERNALITIES

Whatever has gone on in the real estate market has had significance for neighbors. No one has been "an island unto himself." Externalities have become a topic of wide interest among economists, and those studying land have typically made explicit reference to externalities.

Private operators and government have both helped or hurt non-participants in the transactions. The quality of life has depended in many ways on what others did. Many positive results could not be included in the rewards available to those who do the job; such good consequences have entered into the calculations of those who decided whether or not to make the effort. Adverse consequences to others, such as air and water pollution, have often escaped the calculations of those who made the decisions. If too much good land moves from food production to housing, future Americans will perforce face more expense for food as against shelter.

The inadequacies of private markets in dealing with externalities have been widely noted. Our concepts have lacked precision. Scope for vagueness was large, and trying to offset market failure produced errors. Be that as it may, current and future decisions seeking optimum patterns of land use should try to make the best of externalities, encouraging the positive and discouraging the negative.

Inherently, concern for third party interests has almost always suffered from gaps in knowledge and differences in bargaining positions. Citing externalities to justify nonmarket action has tempted (perhaps unconsciously) bureaucrats to grasp more power than has later seemed desirable. Both private operators and government have often assumed that right and justice warranted riding over the interests of third parties. Where compensation did not have to be paid and where beneficiaries could profit without bearing the expense, politically motivated actions have often lost sight of goals which were reasonable in the broad sense of public interest.

The private market has often neglected elements resulting from their choices. Some parties were not required to bear costs of benefits for which they were in fact responsible; others were not compensated for burdens they were compelled to bear. The range was from amounts that were large, relative to the concern at hand, to matters that were identifiable but *de minimis*.

Reference to externalities calls for some awareness of alternatives, especially when the issue is as broad as land use. But taking even major alternatives into account could be very difficult, and many intangibles are hard to identify.

Government could require respect for considerations other than those representing the views of the marketplace. Government could consider the interests of parties who are affected but lack legal privity, such as persons living downwind from a smokestack. A city council, board of county commissioners, or a bureaucratic agency could seek out and examine the significance of externalities. Regulations have often reflected enlightened compromise improving greatly upon results otherwise attainable.

One merit traditionally claimed for the free markets was that they revealed continually what people wanted and how intensely they wanted it, but externalities were only sporadically appraised in such markets. Therefore, the information needed for accurate recognition of externalities of cost and benefit was mostly inadequate. But nonmarket devices, such as public hearings, sometimes have gotten near enough to adequacy to permit conclusions which were clearly better than those of the free market. Every political process has swayed in the winds of politics, and many have led to actions which in no way improved upon the free market. Decision makers did not have to use their own resources and were sometimes fatuously open-handed. Politics and bureaucracy have earned less than the highest of repute as agencies for successful, efficient decision making. Farm programs as they have actually operated have given rise to proposals for expansion of governmental determination of "who gets what and why" in the use of land. "Government failure" has a place with "market failure" in any honest appraisal of what has taken place, but neither warrants sweeping generalization.

When considering alternative land uses, the information required by governments or private owners for adequate recognition of externalities, such as environmental elements, has often been varied and complex, and the best available information was often imprecise. Government actions have too often failed to weigh the costs and benefits of keeping old cities safe and attractive by effective policing and tidy municipal housekeeping compared with the building of new cities on peripheral farmland.

The change in the cost of oil altered many calculations involving location and land use. Private markets have shown great capacity to adjust to what is known. Even so, government policy has shifted from time to time, and private participants had trouble keeping up. Private markets could not walk a straight course when government was staggering uncertainly.

ECONOMIC AND TECHNICAL FLEXIBILITY

The fixity of the earth's area did not affect all elements relevant to land use decisions. Land has become more or less productive. Output per acre has increased enormously in some areas and has dropped to zero as deserts expanded. Erosion destroyed the capacity to produce. Cutting of timber, extraction of minerals, and other changes have affected not only the land immediately involved, but also future production. To varying degrees, technical possibilities exist to slow or prevent crop reductions.

Some actions technically available have been economically desirable in one time and place, but not in others. Costly desalination of sea water to irrigate fields was warranted in Israel, but not in the United States up to 1980. Owners have almost always tried to preserve the worth of their land or to sell to someone who could when the prospects seemed to justify use of resources beyond their scope. "Smart operators" searched out such opportunities.

Yet skepticism about the processes is justified. The former owner's understanding might have been incomplete; financing, through a quirk of the mortgage market, might have been available to the buyer but not the seller; or overeager owners might have been tempted to spend more than justified by the net benefits. The difference, often measured in millions of dollars, might simply have represented the relative degree of sophistication of the parties or, to put it bluntly, "who knew whom and how well."

Transportation facilities and costs have always influenced the usability and worth of land. In a meaningful sense, mankind has enlarged its access to land by improving transportation. Few consumers in the twentieth century have had any reason to think about the many sources of their food. Almost literally the food of the world became available in American supermarkets. Americans no longer depend on nearby farms even for fresh fruits and vegetables. People trade willingness to bear transportation costs for the ability to rely upon food sources close to home, and the market automatically balances these and a host of other considerations.

It was difficult in the late 1970s to imagine that future Americans of high incomes would not be able to draw amply on the world's agricultural output. America could pay enough in transportation, although presumably transport expense would rise with energy prices.

But would the market in the 1980s, in deciding upon shifts of land near cities, take adequate account of the worth of proximity if transportation costs should rise drastically, or would this prove to be a future imponderable beyond the scope of the market to rationalize? This is the kind of imponderable in connection with which free markets have often wrongly combined incentive and knowledge. Occupation of flood land near the Mississippi River is such a case. Cultivation of the slopes of Mount Vesuvius, deplored by the Roman Pliny, is another; and construction of buildings near the beds of "dry" rivers in Phoenix, Arizona, a third. The issue is easier to be raised than to be resolved with confidence in any answer. The potentials of transport are enormous, but disruptions prove to be alarmingly possible. As incomes in other parts of the world have risen, effective competition for top-grade agricultural output increased also.

Discussion of future food supplies has implied that domestic and close-in sources were to be preferred for considerations of national defense. Foreign countries facing threats felt that free market decisions could not be counted on to provide for developments that might grow out of unexpected political and military changes, and they preferred to head off such contingencies by public action to keep land in good production.

Rising populations provided past incentives for changing land use. New uses were found when transportation and other cost-to-benefit ratios justified. Distant but fertile land grew food while close-in land went to sprawled housing, but this was far from a "free market" decision; federal intervention greatly influenced the free mortgage market.

RESEARCH

The fruits of land have never been fixed, and scientific advances have had significant impacts. But the private market could not be relied on to develop fully the potentials of advance knowledge or adopt them in an orderly sequence. The pell-mell shift from hand labor to big machines in the 1940s and 1950s changed the whole system of raising cotton; this threw about twelve million field hands out of work and unleashed social problems that dominated American thinking, especially in the Northeast, for thirty years and more.

The basic units of food production (farms) were generally small relative to the cost of research. Tremendous improvements in land productivity, measured by output per acre, have resulted from a combination of sources, including research of many kinds on many fronts. Universities and publicly financed experimental stations have played a part; so have private manufacturers of processed and semiprocessed foods and of chemicals and machinery. But mechanisms for mobilizing the interests of individual farmers were too weak to assure that they would quickly adapt to serve their best interests.

Some advances in knowledge could be .patented or copyrighted to assure a commercial sponsor of research economic rewards commensurate with the risks. The free market did not devote enough private resources to research to reap the full total of potential benefits.

The potential of land as a source of food for the world's growing population depended in 1980 upon advances in technology and, to an even greater extent, on politics. Areas like the Mekong Valley, potentially very productive, have been drastically restricted by a seemingly endless war, and by early 1980 this war was dragging into its fourth decade. The owners of small plots in "safe" countries lacked resources and incentive to finance research. Owners in war-torn countries were without political power to end the wars. Independent and public foundations and research organizations have undertaken research. But by 1980 no organization had found a way to stem the ambitions of greedy dictators, to cool tribal rivalries, or in any way bring peace where the swords clashed. Any history of the European peasantry from Charlemagne to 1980 makes terribly gloomy reading. The peasants' sons were conscripted and killed, their daughters deflowered, and their crops trampled. The hooves of the war horses never thundered gloriously, but were always destructive.

TAXATION

Property taxes must be paid, each year, in cash, and income taxes have complex relations to real estate. The standard rule of property taxation in America has been that the tax was fixed to capital value rather than to income actually realized. On the urban fringe, land values have been substantially greater than justified by capitalization of income from current operations. The discrepancy does not conform to the popular but somewhat vague concept that property taxation should relate at least vaguely to income. Gains in the form of unrealized appreciation do not provide cash to pay yearly taxes except as a rising base for borrowing.

Income taxes and death and gift taxes have many features which affect land use. The owner-occupant of a house can deduct interest and real estate taxes from the base on which he pays his income tax. The owner of rental housing can deduct these and also depreciation. Elements of tax law require sophisticated knowledge but greatly influence private investment in real estate; appropriate investment in real estate can shelter income from the rigors of high income taxes. Market prices for land have for generations reflected tax factors which did not grow out of the inherent elements of supply and demand. The net result was to enlarge the demand for land compared with other investment and set in motion "artificial" forces influencing the times and terms of buying and selling.

The market economy of the United States has operated in a framework in which local governments pay for their operations largely through property taxes on capital value of real property. One result has been pressure to put land to its "highest and best" use. Market forces have indicated a certain pattern of land use as potentially best, recognizing all the discontinuities and "imperfections" of real estate markets. Tax assessors have been generally supposed to use such values until the special usufruct assessment laws were commonly applied to farms and woodlots in the 1950s and 1960s.

Economists have often said that the owner's best interests would lead him to utilize his land to best advantage. Taxation that rests on current capital value was thought to add only slightly to the incentives to get maximum income. Realities, however, show the existence of different pressures. The need to pay the property tax annually in cash adds pressure to convert earlier than might otherwise occur, since it reduces the small owner's ability to wait.

Market conditions were changed by usufruct assessment practices in most states which favored retention of agriculture. Nevertheless, free market consequences were less clear than they might have been. The tax reduction on farmland enhanced market value, brought profitable liquidation closer to the farmer, and made it easier for a speculator to "hold on." The results varied from case to case and community to community depending, in part, upon what happened to tax-financed services, to tax rates on total assessments, and on how much the tax relief consisted of postponement as distinguished from permanent escape.

GOVERNMENT SPENDING

Highway construction, street widening, irrigation and water projects, sewer building, rivers and harbors legislation, and many other normal aspects of government spending have affected the market for land in general and for some parcels in particular. Housing subsidies, farm aid, and miscellaneous urban betterment programs also influenced the market. The endless catalog of government spending at all levels shows example after example of actions which altered the conditions of land use. But political changes which might be nationally wise in any long run had to be undertaken in local areas where decisions have usually rested on local benefits in the very short run. In most cases, the public, other than the direct users and owners, had little or no concept of total results.

Future Adjustment Under Conditions of Increasing Strain

Looking to the future from the background of 1980, population growth may be relatively modest in this country. Persons now living may not see increases much beyond the eighty million added since World War II. World-wide, however, population increase is projected in the billions, and untold multitudes will likely look to American supplies of food. American economists can only guess how market forces will operate if demands on tillable land rise substantially, available acreage shrinks slowly, and technology cannot assure improved output per acre.

People have managed to survive by adjusting to what they had to. Unpleasant as well as pleasant changes were taken in stride. Decisions were made in light of alternatives available at the time of decision. The changes in America's way of living have within living memory shown some of the powers of adjustment. Much change in America during this time has been "upward" and thus appealing; but some changes, like the tragic deterioration of security on streets, have been unwelcome.

The food production of 1980 seems to portend strain which could take the form of rising prices appearing gradually, a little here and a little more there, all superimposed on the normal changes of the market. Some of the adjustment would restrict price declines which productivity increases would otherwise permit; most of such effects would be hard to identify. In an economy where the price *level* changes due to inflation, food prices would presumably go up more than most (except energy), certainly more than if the supply of food-producing land had not been reduced. The stresses and strains on society, if they should happen, can be easily predicted. But many other developments, mixed in constantly changing ways, would also be occurring, and the cumulated results would take forms not presently foreseeable.

Conditions developing from year to year would probably be those of an interconnected world. Despite many obstructions between 1945 and 1980, supplies moved from one place to another to meet demand and would probably continue to do so, although various kinds of government intervention would influence results. As pressures on food supplies increase the world over, relative price rises can be expected to affect other parts of the world relatively more than North America.

Any past or future action taken in this country to preserve land for farming would presumably benefit consumers the world over. Government is unlikely to restrict American agricultural exports within the near decade. The political strength of agriculture will remain formidable and can be used to prevent limitation on selling in markets abroad. An "eaters' bloc" might mobilize to keep food in this country despite the constitutional prohibition of export taxes, but such a prospect appears unlikely.

Through much of history, agricultural exports have been important sources of funds to pay for imports. In 1980, farm exports paid much of the rising bill for imported petroleum. Any substantial reduction in the area of cultivable land would diminish export capabilities of farm products. With flexible exchange rates, any adjustment to reduced exports would be gradual but painful. Import costs would rise, but if a shift of land use permitted an increase in other forms of productive capacity, the net effect on exports might be actually favorable. But substantial growth of manufacturing need not necessarily take land from farming.

Government versus Private Markets

No really "private" market for land has existed in the West for centuries, if ever, and it never worked the way it was designed. Centuries ago in England the barons acquired vested interests, and during the reign of King John they extracted Magna Carta from the Crown. This launched the movement to greater freedom in the holding of land.

The feudal and manorial system eventually vanished, but it took a long time. The evolution of feudalism into modern land markets is a fascinating story but too long to tell here. By 1900, its last traces were pretty well erased except for land entailments. America never had anything like feudalism, although the southern plantation system had recognizable similarities. In America, as in England, the market was probably at its freest around the turn of the century. Then followed a long eighty-year period of increasing restriction.

The first new interference with a free market came in the form of zoning laws. The next, historically, was the Federal Land Bank System, which changed the pattern of farm mortgage financing. Other forms of interference followed, with a rash of public actions in the 1930s. Significant among these actions was an effective government action to prevent farms from going through wholesale bankruptcy and emerging at a price level adjusted to then current conditions. This had happened before in 1837 and to some extent in 1873. The actions of the 1930s eased conditions for individual farmers but actually saved innumerable financial institutions which were heavily invested in farm mortgages. Parallel actions directed to head off mass foreclosures of residential and commercial mortgages involved the Home Owners Loan Corporation and the Reconstruction Finance Corporation.

Then in the 1930s, the Federal Housing Administration, heavily involved in financing urban development, contributed to urban sprawl. Meanwhile, zoning and building laws became increasingly common; much suburban zoning regulation involved very local considerations, and much of it was contrary to the best metropolitan interests. Finally, in the 1970s, environmental legislation further restricted the market and made development less a matter of the owner's right and more a matter of government privilege to be bestowed or withheld by a complex of boards.

Each of the twentieth century restrictions on land use was designed to remedy some defect in the performance of the private market. The point is that America has no recent history of a really free real estate market, and modern economists cannot judge in the light of experience exactly what a free market would have done differently from what happened. Certainly the public restriction of further conversion of farmland to urban uses would be in keeping with the trends of the twentieth century rather than the opposite.

Economists can estimate the relative effectiveness of decision making by public bodies and by private markets. Private markets have made decisions faster and generally seem to have evaluated short-run considerations as well as government. Long-run considerations have been different. Local governments have seldom seen very far ahead and have made quite parochial decisions reflecting short-run interests important only to the locality immediately affected. State governments and the federal government have taken a longer view.

The decision whether or not to preserve farmland involves the "long run," however defined. Land, as repeatedly stated, is fixed in supply, and shifts from farming to urban uses are irreversible. The indications are that the private market made decisions in the past, and presumably will make them in the future, which will involve conversion of too much farmland in the wrong places. Cities, according to the thinking of many planners in 1980, became too sprawled and need a period of recompaction. Threats, real and imagined, to the transportation system suggest that too many people are at too far a distance from food supplies for safety in a world that looks more hazardous with each passing day.

The conclusion seems warranted that where it would work, the private market is to be preferred; but that in deciding a question of farm preservation, public action is required.

Mark B. Lapping

8

Agricultural Land Retention:

Responses, American and Foreign

The American Perspective

THE NATURE OF URBAN/FARM CONFLICT

The conversion of farmland in the urban/rural fringe into other more intensive uses was stimulated by a number of factors inherent in the nature of American life in the post-World War II period. The construction of highway systems well into the hinterlands, subsidies and tax deductions provided for single-family dwellings, the rise of tax-exempt municipal bonding, utility rate structures which promoted service extensions, and cheap and available energy resources have all conspired with a vision of the "middle landscape" as the ideal life style to bring millions of acres under the bulldozer and pavement. While the obvious result was a shift of enormous proportions in the allocation of land uses, this phenomenon also tended to undermine

MARK B. LAPPING *is associate director of the Environmental Program and an associate professor of Environmental Studies and of Natural Resources at the University of Vermont. Dr. Lapping has served as a consultant to regional planning commissions and to several government agencies, including the National Agricultural Lands Study. He has authored numerous monographs and articles on land use. Dr. Lapping acknowledges the aid of Cynthia White-head of the Conservation Foundation in Washington, D.C.*

the viability of agriculture. The National Association of Conservation Districts estimated that between 1967 and 1977 almost thirty million acres of farmland had been converted to urban uses, a rate of nearly three million acres per year. The majority of the farmland conversion to other uses occurred in urban fringe areas, and this was all the more serious because much of that land was of very high quality in terms of cropland productivity.

In the broadest sense, the conversion of farmland introduced a sense of instability and impermanence into these changing regions. Some of the elements of dissonance were direct: traditional farming practices were deemed nuisances and hence were subject to legal conditions or outright prohibition; tax levies upon the land were adjusted upward to pay for newly demanded services; crops, animals, and equipment were destroyed by vandalism, harassment, or the externalities of growth and development. Other impacts were less direct but were fundamentally more serious: farmers lost status in the community power structure; agribusiness support systems dissolved as critical mass eroded and a climate of speculation came to permeate the land market.

Agricultural Land Retention Programs

Emphasis within the United States tended to focus on agricultural land preservation techniques and very seldom on the changing structure of agriculture in an urbanizing environment. These policies reflect the incremental, reactive, and stopgap nature characteristic of land use policy in the United States. With few exceptions leadership in this area has been taken by state and local government with very minimal federal participation.

DIFFERENTIAL ASSESSMENT

With the exception of Georgia and Mississippi, all of the states had, between 1956 and 1980, adopted some form of differential tax program to aid farmers. (The first state to inaugurate such a program was Maryland in 1956, followed soon after by New Jersey.) As a consequence, it is the most common agricultural land retention scheme in the United States. All such systems share one basic characteristic: lands in agricultural use are assessed for their use value as opposed to assessments at market value. Assuming that the use value of the farm is less than the fair market value in developing areas, land assessments, and therefore taxes, should be lower for farmers. The expectation is that a lower property tax burden would encourage farmers to keep their land in agricultural use, even in the presence of conversion pressures.

Basic Types of Differential Assessment—Differential assessment approaches are of three basic types: preferential assessments, deferred taxations, and restrictive agreements.

Under a *preferential assessment* law, land in agricultural use is assessed on the basis of its value in that use. No penalty is imposed on the landowners if the land is put into another use at some later time. Seventeen states utilize such a system: Arizona, Arkansas, Colorado, Connecticut, Delaware, Florida, Idaho, Indiana, Iowa, Louisiana, Missouri, New Mexico, North Dakota, Oklahoma, South Dakota, West Virginia, and Wyoming.

Under the *deferred taxation* strategy, land is assessed according to its value as farmland. If the owner changes the use of his land to some use which does not qualify under the law, a deferred tax, or a rollback, is levied. The amount of the rollback is equivalent to the tax savings received by the owner for a designated period of years preceding the change in land use. Twenty-six states have this type of program: Alabama, Alaska, Illinois, Kansas, Kentucky, Maine, Maryland, Massachusetts, Minnesota, Montana, Nebraska, Nevada, New Jersey, New York, North Carolina, Ohio, Oregon, Pennsylvania, Rhode Island, South Carolina, Tennessee, Texas, Utah, Vermont, Virginia, and Washington.

A state or local government might make *restrictive agreements* with landowners by which farmers agree to restrict the use of their land to agricultural purposes for a specified period of time in exchange for use value assessments. After this period of time has elapsed, the landowner is free to alter the use of his land without rollback or penalty. Six states, California, Hawaii, Michigan, New Hampshire, Pennsylvania, and Wisconsin, have such programs.

These three systems have two basic differences. First, those revenues which the landowner saves under a tax deferral system, or a portion of them, would be recovered by the appropriate level of government when the land is converted to nonqualifying uses. This recovery technique simply does not exist under the preferential assessment system. Second, a willing government has to enter into a restrictive agreement and has to first establish that an agreement with a landowner constitutes a gain for the public. Under preferential assessment and deferred taxation approaches, a landowner is entitled to participate simply by virtue of the fact that his land is eligible as defined by statute.

Other Mechanisms for Differential Assessment—Several other mechanisms have also been created to provide less than fair market value assessments for farmlands.

Under the *classified property tax,* different assessment ratios are applied to different types of properties and land uses. The assessment ratio is fixed for each type of land use. An increase in market value results in equal percentage increases in the assessed values. Most use value assessment programs do not provide for market fluctuations since no predetermined assessment ratio is established.

Under the *general direction approach,* the local assessor is directed to consider all current land use controls to be permanent. No potential for development exists under such a program, and farmlands, which are assumed to be zoned for agricultural or open spaces uses, cannot be taxed at suburban values. This system freezes uses into their zoning designations, and development potentials are ignored.

In order to provide *exemptions from or limitations to applicable tax rates,* statutes which exempt farm structures from the property tax could be broadened to include the land itself. Alternatively, the property tax could be applied in a limited fashion, with a lower tax rate on farmland than that applied to other types of real property. Such an approach is possible under Iowa law.

The *"circuit breaker" method* stipulates that payments which exceed a certain percentage of income will be deducted from state income taxes or directly rebated to the farmer. A serious problem with this program is that possible rebates to speculators might be greater than those available to farmers. Michigan attempted to address this concern by requiring that the landowner forfeit his rights to develop his land in return for circuit breaker relief. Wisconsin also adopted this approach, and the legislatures of North Dakota and Minnesota were studying bills in 1980 to create such a program.

Effectiveness of Differential Assessment Programs—Differential assessment programs, whatever their nature, proved not to be fully effective in preserving agricultural lands. Studies of the California program, commonly known as the Williamson Act, uniformly conclude that the potential effectiveness of use value assessment as a means of reducing both premature land conversion and urban sprawl was minimal at best. New Jersey was losing three farms a day in the 1960s. After differential assessments were introduced, the rate slowed to one every third day in the 1970s. Studies of programs in other states consistently came to similar conclusions.

For a number of reasons these programs are not fully effective. Perhaps the most obvious reason is that owners of land near population centers, who are most susceptible to development and conversion pressures, are noticeably unwilling to participate in differential tax assessment programs, especially those which have significant penalties for conversion of land to nonqualifying uses. Yet judging the effectiveness of programs by high levels of participation might be misleading since high enrollments of otherwise developable land might encourage a leapfrog type of development which makes sprawl worse and increases the cost of providing necessary public services.

Perhaps the more significant reason why such programs tend to be less than effective is that tax concessions are largely overshadowed by the opportunities associated with development. This is the result of two factors operating within the land market. Penalties imposed by rollbacks are usually quite small relative to development opportunities. The size of the penalty depends on the divergence of market value from use value, and the larger the potential rollback penalty, the larger the potential capital gain associated with land use conversion. In addition, rollback taxes or other deferred taxes are deductible for federal income tax purposes. The amount of these penalties is frequently less than the amount of the tax which the owner has saved. Thus, he will seldom pay more taxes than if he had not been in the program, and he will often pay considerably less. If there is no interest penalty on the deferred tax, which is often the case, he will also have gained the free use of the money involved in his transactions.

Assessment of high-quality agricultural land at higher use values than poor land makes the conversion of better farmland more likely, since it is subject to smaller deferred taxes. A farmer planning to sell all or part of his land upon retirement might decide to sell his land earlier instead of seeking tax relief since the accrued tax obligation would lower its market value.

Preferential assessment appears to be a poor system to influence land use, perhaps the least effective means of farmland preservation. It imposes no responsibility on the landowner to maintain a working agricultural use of the land, and since it is available to any landowner who meets the area requirements, it tends to frustrate broad planning procedures. Several states, such as Maryland and Connecticut, found such strategies unsatisfactory and moved toward either deferred taxation or restrictive agreements, or other programs entirely.

Tax deferral systems remove one immediate and positive incentive to convert land to nonagricultural uses, but the only real advantage to the farmer is a postponement of taxes. If the program were only voluntary, landowners might not enroll in it, fearing an accumulation of deferred taxes. In addition, a tax deferral approach is somewhat expensive to establish and administer.

It is not entirely clear whether the rollback tax discourages land speculation, though it probably has little significant effect. Any real hope for the prevention of land use conversions depends on the participating farmer being penalized severely enough in the form of a lower sales price so that he is discouraged from selling. Only in this regard could the rollback encourage, and then not really guarantee, farmland retention. Certainly the rollback tax has some marginal effect upon land use. More importantly, however, it provides society with a means to recapture tax concessions which have been made without securing the intended social benefit.

Restrictive agreements, on the other hand, seem more likely to accomplish the objective of influencing land use. Such agreements penalize a change in use, and they are closely related to planning objectives. Yet participation appears to be limited to those who are truly confirmed in a farming future.

Differential assessment alone could not be expected to curtail development pressure when the market value of farmland greatly exceeds its use value. A conversion to a nonagricultural use is irresistibly profitable. Use value assessments combined with a commitment to farming or immobilities in the farm labor and capital markets influence landowner behavior to the extent of eliminating the immediate need to convert. Some commentators suggest that the system could be improved if use value assessment were mandatory for all land under qualifying uses. It might also prove effective if complemented by well-planned agricultural zoning to reduce developmental expectations, but this raises other issues related to concepts of justice and equity.

Some Criticisms of Differential Assessment Programs—Differential assessment could reduce the tax base of a jurisdiction and thereby reduce local government revenues. The ability of a county or town to supply certain services might be sharply curtailed if a significant amount of land were brought into the program and taxes on non-participating land could not be increased to compensate for such losses.

One study in Maryland determined that eight counties on the fringes of Baltimore and Washington, D.C., saw a reduction in revenues of approximately 3.6 percent. In an effort to address this problem, both California and Vermont reimburse local taxing jurisdictions for revenues lost from use value assessment. Missouri also makes compensatory payments, but only in lieu of taxes forgone by differential assessment on forest plantations.

Differential assessment raises the property tax burden on nonfarm landowners. This effect is reduced under deferred tax schemes, since the burden shifts as land is changed from qualifying to nonqualifying uses. It has been argued by some that it is one thing to help farmers but another to aid speculators. Since many differential assessment programs do not distinguish between bona fide farmers and specu-lators, they are sometimes described as a subsidized license to speculate. Speculators might often benefit and be encouraged to hold out for greater future profit at the cost of those who have to absorb property tax shifts.

It is also argued that differential assessment programs which apply on a statewide basis unnecessarily benefit strictly rural land beyond the pressures of urban development by treating it equally with treatment to those farmlands directly within the shadow of urbanization. This argument overlooks the fact that market value is closer to use value for farmlands beyond the pressures of urban development than is the case on the urban fringe.

Conclusion—The key policy issue which emerges from discussion of differential assessment is who would pay for the benefits perceived from land use control. Land use controls aimed at the preservation of open space amenities seem to generate widely dispersed and generally small increments of benefit. The direct costs, on the other hand, might be highly concentrated upon the few owners whose land use options are limited, the local government that loses a property tax base, and the region where development is limited.

Differential assessment laws could be justified on the basis of promoting tax equity if the laws were so written that only the intended persons were eligible to receive benefits. This would appear necessary since a number of studies suggest that the biggest beneficiaries of such programs in the 1970s were large landowners rather than family farmers.

CAPITAL GAINS TAXATION

Another process which could influence farmland conversion is the capital gains tax, distinct from the federal capital gains tax and levied by a state. This concept was first articulated by Henry George in his classic *Progress and Poverty* (1870). Though a number of states were discussing such a program in 1980, only Vermont has implemented one. Under Vermont's program, variations in the tax rate depend on both the degree of gain from a land transaction and the length of time the land is held prior to sale. The rates rise as the percentage of gain from a sale increases, but decrease over time. Long-time owners are taxed far less than short-term, fast turnover owners. The highest tax rate is 60 percent of the gain on land sold during the first year after purchase where the gain is 200 percent or more and down to 30 percent if the gain is 100 percent or less. As each year passes after the initial purchase, the rate drops in each gain class, until the sixth year when the tax is eliminated altogether. The purpose of this law is to tax the capital gains on land so heavily that speculation and the rapid turnover of land lose much of their profitability. A number of evaluations of this program suggest that the results are mixed at best. Where land markets soften, it would seem that the capital gains tax has been of aid in slowing the loss of agricultural lands. But where the market is intense, sales remain brisk.

ZONING

Without doubt, the best known program to influence land use is zoning. The constitutionality of zoning as a legitimate exercise of the local police power was established in 1929 in various cases such as *Village of Euclid v. Ambler Realty Co.* But government cannot use zoning to impair a landowner's rights unreasonably, purely on the basis of policy.

The great difficulty in relying upon zoning to preserve agricultural land use is the lack of objective standards for determining whether or not property is being restricted in a reasonable way. While zoning regulation need not permit the most profitable use of the land, it is also true that regulations often are of little benefit to the farmer whose land is zoned for agriculture. The farmer has to have the right to farm. When a regulation imposes serious burdens without any compensating benefits, the regulation might be held an unconstitutional "taking."

Minimum Lot Size Regulation—One common approach to zoning to preserve farmland is the adoption of minimum lot size regulation. There has been a trend toward invalidating large lot zoning in suburban areas on the grounds that minimum lot size requirements unreasonably exclude the poor and minorities. Much of this activity took place in New Jersey, and the leading case was *So. Burlington County NAACP v. Township of Mount Laurel* (1972). However, where carefully designed so as not to create a discriminatory effect, the courts are recognizing the value of such minimum lot size regulations.

Exclusive Agricultural Zoning—Under this other regulatory method, only farming is allowed in areas zoned for agricultural uses. This benefits farmers in terms of tax rates since market value and use value of the land tend to converge. The possibility remains that such an ordinance could be struck down if challenged on the ground that it is confiscatory since it prohibits any land use except agriculture, and land in the zoned area not suitable for farming would be zoned into economic uselessness.

Through enabling legislation, at least twenty-one states reserved for local governments the right to zone land for agricultural use. Such programs did not, as a rule, prove to be very popular. Instead, many of these same states found it more advisable to permit other land uses in predominantly agricultural zones. This type of nonexclusive zoning apparently resulted from significant pressure from the farming sector. The consequence was that this technique was not very effective in preserving agricultural land uses. Hawaii remained in 1980 the only jurisdiction where exclusive agricultural zoning is effective. All of the state's land is zoned in either urban, rural, agricultural, or conservation districts. The results which Hawaii achieved are largely the consequence of a unique State Land Use Plan, a different cultural tradition toward land usage, and elements peculiar to tropical agriculture.

Compensable Zoning—Although this third method involves more than simple regulation, it has been held a valid exercise both of the police power and of eminent domain. Under this system an owner is compensated for the loss of development value of his property due to the restrictions placed upon his land under the police power. Compensation is based upon development value at the time the restriction is imposed for open space or agricultural purposes. The owner receives compensation only at the time he sells his property, the rationale being that he incurs no loss prior to a sale. Condemnation costs are paid by assessing those within the planning district who benefit by the regulation. The benefit assessment allows a recapture of some of the (planning) benefits and their redistribution to those who suffer injury. Administrative problems together with costs of condemnation make this program expensive, prohibitively so for some areas. However, the concept of compensable regulation is fundamental to the purchase and transfer of developments which will be discussed later.

The most significant weakness of zoning as an agricultural land retention device, constitutional problems aside, is that even the most carefully prepared zoning map and ordinance are subject to variances, zoning amendments, and special exceptions. Alterations in the structure of zoning sometimes tend to overwhelm ordinances; they are "nickel and dimed" to death by exceptional cases. Moreover, as development pressures approach an area, speculators who anticipate the granting of variances cause increases in market value and assessed value, which consequently increase property tax burdens on all landowners. Another drawback is that zoning decisions are normally made by politically vulnerable local governments. Finally, the lack of coordination of policies among different political jurisdictions with conflicting interests and the absence of mandatory state guidelines or supervision could easily frustrate comprehensive regional planning.

The integration of zoning with use value assessment could influence significantly the timing of development. All land within a specified area could be zoned exclusively for agricultural use and could be designated by the county as an agricultural preserve, thus making all land eligible for use value taxation. In this context, zoning becomes less vulnerable to the "unfair taxation" argument. Both California and Oregon have integrated their zoning enabling legislation with differential assessment laws. Oregon's approach also utilizes a statewide Land Use and Development Commission which requires that all cities and counties adopt comprehensive land use plans. One of the requirements of this program is that cities must designate an urban growth boundary around urban areas. The dimensions of boundary lines must be such as to preserve as much farmland as possible. Nonurban counties are permitted to establish exclusive agricultural zones, and any subdivision of land below ten acres must receive agency approval.

FEE SIMPLE PURCHASE AND LEASEBACK

It is argued that governmental authorities should simply purchase farmland under the power of eminent domain and then rent the land back to farmers. As a farmland retention method, fee simple purchase and leaseback did not have extensive application in the United States. The Canadian Province of Saskatchewan had over one-half million acres in such a program. This was part of a larger rural development scheme which the province was developing. The success of this venture might be attributable to the number of small farmers in the province. Applicants for land could not have had annual net incomes averaging over $10,000 in the prior three-year period, nor could an applicant's net worth during this period have exceeded $60,000.

Purchase and leaseback arrangements might benefit the farmer by reducing estate taxes and property taxes. The farmer receives full compensation for the land he sells, but the loss in pride of ownership is significant. The farmer becomes a rent-paying tenant instead of the owner-operator, and as a tenant he is less likely to make the substantial investments in the land or machinery which are often required for truly viable, long-term agricultural productivity.

Governmental agencies traditionally make poor landlords as they are notoriously slow to make decisions about such things as repairs and capital improvement. It is also conceivable that public pressures upon government might create an atmosphere inconsistent with the solid farming practices and policies. Finally, the cost of such a program could be very large.

PURCHASE OF FARMLAND DEVELOPMENT RIGHTS

An alternative to purchasing land outright is the purchase of rights to develop the land. Often referred to as scenic or conservation easements, the concept of development rights purchase generated considerable interest in the late 1970s, especially in those areas where urban conversion pressures were great. Under this method a farmer sells to government his right to develop his land. The farmer keeps the fee interest except for this one right, retaining all of his other rights, including the right of possession. The encumbrance runs with the land and thus binds all subsequent purchasers as well. This method is very much like the sale of mineral rights by a farmer who retains fee simple to the surface. Mineral rights were widely sold in the 1880s and before, Pennsylvania being a prime early example.

The public acquisition of development rights presents several clear advantages over other methods. Buying development rights is likely to be cheaper than fee simple purchase, and the landowner still retains ownership and control of his farmland. Landowners receive a cash payment for giving up their right to develop, and this capital could be utilized further to enhance the farm operation. Following a favorable ruling from the Internal Revenue Service, government could pay farmers for the development rights over a number of years rather than all at once, spreading the federal capital gains tax and keeping it in manageable proportions when it has to be paid. Farmers who sell their development rights still have to keep their lands on the tax rolls, but when the development rights are transferred to government, the farm can be taxed only on its agricultural use value. Moreover, the expense of maintaining the property is not transferred to a governmental agency, and the farmer can continue to keep his property in a productive, though limited, use. A more subtle, though very important, attribute of the program is that it addresses in a direct way some of the equity issues related to farmland retention.

All too often a farmer's land is at once his hospitalization plan, insurance policy, child's college tuition, and personal retirement fund. Consequently, farmers are clearly concerned about the issue of compensation when land use controls are established that they perceive as limiting their options. When compensation is provided, as in the development rights purchase, members of the agricultural community are more likely to participate. (Wisconsin's new system ties retention to a reduction in income tax liabilities.) The point is that techniques to keep farmland in agricultural use often require that the development potential for agricultural lands will no longer be available to farmers. For a program to succeed, it clearly has to have the endorsement of the farming community. One element in obtaining such support and cooperation is dealing with the issues of justice and equity openly and directly.

The big problem of buying development rights is cost; the purchase of such rights can be expensive. In rural areas, where development pressures are less intense, the cost of buying development rights is lower. But in areas under conversion pressure, the cost of acquisition would be high, and the reduction in the local tax base is significant as lands have to be assessed at use value. One way to reduce such costs might be to target such purchases selectively so that only those farms under direct development pressure would have their development rights purchased. Still, funding is required for such a program, and it appears that bonding, either by local or state governments, is the most appropriate source of capital for the program.

In the late 1970s, New York's Suffolk County, situated on Long Island and within commuting distance of New York City, had the most experience with the public purchase of development rights. The program was adopted in the mid-1970s in response to concern about potential overpopulation, the loss of prime agricultural lands, and the diminution of open space resources. The program attracted consistent support from county residents. Some perceived it as a way to guarantee open space; some saw it as a means to control population; and still others thought it would assure plentiful quantities of fresh vegetables and fruits for their tables. The greatest support came from the farming community. This was not surprising, since farmers received upwards of $6,000 per acre for their development rights while still keeping title to their property. By 1977, development rights to some 3,883 acres of farmland had been transferred to Suffolk County at a cost of approximately twenty-one million dollars. At the time of writing, it was anticipated that the program would expand to a total of 15,000 acres (just under a quarter of all farmlands in the county) at a total cost of ninety million dollars.

After a demonstration program in New Jersey, programs to purchase farmland development rights were implemented in Connecticut and Massachusetts, and in 1980 New Hampshire was in the process of buying its first development rights. In these three New England states farmland had been lost at staggering rates. By 1970, Connecticut had only 17 percent of its land in agriculture; Massachusetts, 14 percent; and New Hampshire, just 11 percent. Much of the land which formerly was in agricultural use had been urbanized, since population growth and development in these states was intense. Financing for these programs comes from state bonds. While these three state programs are in their infancy and hence cannot be fully evaluated, it appears that there is wide-scale support for them.

Maryland also introduced a "purchase of development rights" program tied to a system of agricultural districting or the creation of agricultural reservations. Funding for this system was appropriated and was to come from a reallocation of state "open space acquisition" funds and from revenues derived from the rollback penalty which was part of the state's differential assessment program. Funding is limited and only time will tell whether or not the program will be successful.

TRANSFER OF FARMLAND DEVELOPMENT RIGHTS

A possible way around the expense of purchasing development rights is outlined in a related concept known as the transfer of development rights. Under this type of program, a given area is designated as a *preservation* district which is to be kept in agriculture free of any other development. Other lands within development districts might be developed at densities greater than permitted under existing zoning so as to absorb the growth which is deflected from the preservation zone.

Growth within the development district can continue along the lines and at the densities established in the zoning ordinance, but development can exceed that permitted by the ordinance if development rights are purchased from landowners in a preservation zone who have exchanged development rights in the development area in return for the loss of development rights on their land in the preservation area.

Under any transferable development rights (TDR) proposal, the development rights are freely transferable among private parties or between a private party and a public agency at market prices, and the use of these rights is limited only by the comprehensive land use plan for that particular area. A landowner in a preservation district finds himself owning land with fewer use alternatives and owning development rights which relate to lands he does not own.

The TDR system does not fit precisely into the definition of either the police power or the power of eminent domain, and much of the literature generated on TDRs is concerned with the question of whether or not it constitutes a constitutionally valid method of land use management. Such programs have been utilized effectively in Britain for over twenty-five years, but experience in the United States is, at the time of writing, limited to efforts at historic preservation.

TDRs seem to present several advantages over other land use control devices. The timing and pattern of development can be controlled, in part, through zoning and manipulation of the availability of public utility services. Funding by the public treasury for land management objectives does not need to be increased since compensation for the curtailment comes through private sector purchases of development rights. Moreover, TDR systems, such as those proposed for New Jersey, Maryland, and New York, correct a serious flaw of the land development process by charging the costs of such growth to private sector participants rather than to the community.

The most serious obstacle to the implementation of a TDR program is the very real possibility that there will be insufficient demand in the private sector for the development rights. Unless owners of land in the preservation zone can find willing buyers for the development rights at a fair price, the value of the rights to the landowner will be severely limited. In a poorly organized or noncompetitive market, the developer is likely to reap large benefits from the development rights sellers, because he can select landowners willing to sell rights at a lower price. Furthermore, the developer can pass on his costs to consumers in higher purchase prices or rents. The presence of a public intermediary agency, acting as something of a broker for a development rights bank, could resolve this problem by controlling the available supply of development rights and by making market information available to prospective sellers. Yet the inability to guarantee justice and equity remains the greatest single drawback to this technique. Moreover, it is a complicated approach, and unless the local planning and management process is sophisticated, there could be substantial problems.

The value of a TDR technique in an agricultural setting remains questionable, since in the United States most testing and development of TDRs have occurred in urban areas. Farmland owners might hesitate to give up control of development rights which they could not use or whose value could not be realized unless a developer decides to purchase them. Large-scale rezoning allowing nonagricultural uses would be necessary in order to obtain an active market. Furthermore, if the development right is taxed as an incorporeal hereditament, like an easement or other appurtenant right, the owner will get tax relief only when a developer decides to purchase it. It appears that the concept will have to be more refined if the TDR program is to become an acceptable alternative to other systems of farmland retention.

LAND TRUSTS

Placing lands to be preserved into a trust is another method of land use control. At least three approaches are proposed: the *private land trust*, the *public land trust*, and the *community land trust*. The private trust is a private, nonprofit corporation which accepts gifts and donations with the objective of holding land in an open or agricultural state. Two models are the Maine Coastal Heritage Trust and the Lake Champlain Islands Trust, which attempt to inventory and control the use of coastal islands. The relative success of this concept which could be applied to agricultural lands depends upon the wealth of private donors whose donations are motivated by philanthropic considerations and encouraged by the deductibility of the value of the gift from the federal income tax base.

The traditional common law doctrine of the public trust does not appear very helpful in terms of agricultural land preservation. The basic public trust theory is that the state holds the public lands of the state in trust for the people of the state. The public trust concept is limited mostly to tidelands and the protection of fishery resources and navigation. It appears ill-suited for farmland since it precludes the use of land for private benefit.

The community land trust might prove more relevant to farmland preservation. This allows a farmer to allocate the development potential of all of his land to a certain portion of it, e.g., 20 percent, if he dedicates the remainder to a community trust. The scheme proposes a voluntary agreement between government and the owner of the farm which would place the entire farm under the zoning regulations standard at the time of the agreement. The agreement would designate a limited area for development at a density that would create a developmental value equal to that of the whole farming unit. Clearly, the program would be feasible only if zoning or other land use controls allow nonfarm uses of the land designated for development. Finally, the agreement would require the remaining farmland to be offered for dedication to a community land trust.

Since the preservation of prime farmlands is the basic objective, the area selected for development should be the least favorable for agricultural use though it should still allow for development at an economically reasonable scale and be consistent with the community master plan. Upon execution of the agreement, the farm owner could sell the developable land at the current market price or hold on to it. He and his heirs would have the right actively to farm the land dedicated to the trust upon payment of a nominal rental fee. Thus, an operating farmer who could sell his development rights would no longer have his capital frozen in his business. Furthermore, he would not be paying real estate taxes on the farmland. Neither would he lose the capital value, nor the potential appreciation of the development area, if he should choose to hold it for future sale. Property taxes on the developable land would, of course, increase as the market value of that land increases. Experience with community trusts is very limited at the time of writing, and hence the long-term feasibility of such a system cannot be determined.

AGRICULTURAL DISTRICTS

While the community land trust remains largely untested as a farmland retention technique, California pioneered a method known as agricultural districting to serve similar purposes. New York State soon followed. Agricultural districts allow farmers to retain their ownership of the land; the program permits farmers voluntarily to form a district and thereby receive several benefits, not the least of which is property tax relief.

The agricultural district program in New York won wide-based support because it emphasizes initiation of control at the local level. The impetus to create such a district came from local farmers who had to apply by submitting a proposal to the county government, although the state of New York can seek to create districts in areas of "unique and irreplaceable" agricultural value. The process of creating a district is a long one, and many local bases of power and authority in rural areas have to be consulted. A proposal has to be approved by the county legislature, the agricultural advisory committee, and town and county planning bodies. A series of hearings provides a process for general public review before a proposal is forwarded to the State Commission of Environmental Conservation, after which it is sent back to the appropriate county legislature for final approval. Participation in the program is strictly voluntary. A proposed district must include at least 500 acres, and the farmers making the proposal must own at least 10 percent of the land to be included. After a district is formed it is affected by the following provisions:

1. Farmers can apply for use value assessments. Later conversion to a noncomplying use subjects the landowner to a rollback levy on tax savings in the prior five-year period.
2. Except for health and safety regulations, local government cannot regulate either farm practices or structures within the districts. This prevents even the regulation of farm odors, a favorite target of suburban residents who reside in transitional areas.
3. State agencies are directed to modify their policies and programs so as to encourage commercial agriculture.
4. Though it does not preclude the state's exercise of eminent domain, the agricultural district law makes it more difficult. Public agencies which might utilize such powers have to examine alternative areas for their projects.
5. The power of local agencies to fund community facilities which might encourage nonfarm development is modified. This is the least appreciated aspect of the New York program, though its control over public investments, such as sewage treatment facilities, water systems, highways, and other utilities, is fundamental to the success of the entire system.
6. Where special service districts are created to tax land for sewage, water, and nonfarm drainage programs, this power is limited to assure low taxes on farmlands.
7. Eight years after a formation, each district has to be reexamined by the county and the state. District boundaries might be modified but only at these eight-year intervals and not before the initial eight-year enrollment period has expired. The state and county have the authority to continue any district indefinitely, regardless of local wishes to the contrary.

By creating a district, local farmers and governmental authorities acknowledge the intention to protect farmlands from development. Given the large degree and intensity of public involvement necessary to establish a district, much of the farmers' discouragement occasioned by potential conversion and speculation is alleviated, and investment in farming is encouraged. This is especially important in New York where the heavy commitment of land and resources essential to efficient dairying adds little to the development value of the land for other purposes. The local problem of a shrinking tax base is partially offset by state reimbursement of as much as 50 percent of lost revenue.

The program is based on the concept that a critical mass of land is necessary for farming, and the minimum area for a district is 500 acres. If a farmer is not in a district because of an acreage limitation or some other factor, he might still take advantage of the tax deferral aspects of the program by entering into an eight-year agreement with the state. The idea of preserving relatively contiguous large blocks of land seems to induce the nonfarming public to accept and believe in the need of a farmland preservation program.

By the middle of 1978, over 300 districts had been formed controlling approximately five million acres, or well over one-third of all farmland in New York. Districting was most popular in rural and semirural areas where the probability of selling farmland at greater than farm values in the foreseeable future was not great. Conversely, in developing or suburban areas the program met with resistance, and few districts were formed within a twenty-five-mile radius of major cities. The reason, of course, is that farmland owners in the urban fringe areas anticipate an imminent conversion of their land to other more intensive uses.

AGRICULTURAL VIABILITY PROGRAMS

As the previous discussion demonstrates, much American policy is oriented to preserving farmlands; the use valuation of land is a fundamental element to almost all programs. Yet it is argued that it makes little sense to preserve farmlands unless there are farmers willing and able to produce the food the nation and world require. Put more simply, farmland without farmers might be irrelevant. Methods to retain farmlands have been successful only to the degree that they were part of a larger effort to enhance the economic viability of agriculture. Several states began to recognize this link and took important steps in the areas of financing and marketing, which are vital to a healthy agricultural sector.

Perhaps the most ambitious strategy is that of Minnesota, articulated in that state's Family Farm Security Act of 1976. Minnesota agriculture suffered from problems common to other areas: tight capital, a shortage of rental lands, and rapidly escalating farmland prices, all combining to create an atmosphere in which young people found entry into commercial agriculture difficult. It is the purpose of Minnesota's Family Farm Security Act to help individuals who cannot secure credit to buy farms and build up equity. In addition to guaranteeing loans to qualifying applicants, the interest payment for a given year can be adjusted because of drought, fire, or other extenuating circumstance. As of early 1979, approximately 110 applications for loans were approved, and the program had over sixteen million dollars outstanding.

In the area of marketing assistance, no state is attempting to do more than Pennsylvania. This state's programs are oriented toward direct marketing, a process which essentially bypasses traditional marketing structures by bringing consumers and producers together. In Pennsylvania this often means that state government acts as a facilitator between producers' cooperatives and farmers' markets. Such programs seem to have the potential of lowering food prices to consumers while enhancing profits for farmers. Moreover, as consumers gain a better understanding of the nature of farming, their concern for farmers turns into political action in support of farmer-oriented legislation. Pennsylvania's attempts at direct marketing resulted in the creation of a broad-based coalition of producers and consumers. Similar efforts at direct marketing are being implemented in New Jersey, Maine, Tennessee, Vermont, and West Virginia. In New Jersey and most other states, roadside markets and "pick-your-own" efforts are successful.

California's approach includes creation of a "hotline" between consumers and producers to bring about better communication between these groups. This is but part of a larger effort known as the California Small Farm Viability Planning Project which advocates several steps it believes necessary to improve financing and marketing conditions for California farmers.

Still some other states, notably those in New England, attempted to coordinate better agricultural and land policy. In 1979 Governor Hugh Gallen of New Hampshire called a Governor's Conference on Food and Agriculture and brought together a variety of people concerned about that state's dwindling agricultural industry. The conference held hearings, issued reports, and made recommendations to advance policy in many areas related to agriculture. One result was the creation of a development rights purchase program. Also in 1979, Maine released a comprehensive report on various aspects of agriculture in Maine— farmland, marketing, transportation, energy use, and finance. Many of the recommendations of that group were addressed by the Maine

Legislature. Massachusetts also adopted a food and agricultural policy and Vermont examined a number of options for agriculture through its economic development planning process. The New England states probably took more aggressive postures in terms of "Yankee" agriculture because their region is at the end of both food and energy pipelines. Nowhere has the loss of agricultural land been as great as in New England, which imports 85 percent of its food from other areas, and at substantially higher prices than elsewhere in the United States.

Still, the vast majority of states has not sought to bring together into a coherent whole efforts in agricultural land preservation and agricultural development. Most states still maintain the official mythology that these concerns can be handled independently of one another. Until greater coordination and support emerge, efforts in both areas will be frustrated.

FEDERAL EFFORTS TO RETAIN FARMLANDS

The federal role in preservation of farmland efforts is minimal. Federal policy has tended to neglect this issue, though agricultural policy is dominated by federal agencies and congressional mandates.

The United States Department of Agriculture (USDA), the principal federal policy-making agency for agriculture in the 1970s, took some potentially more progressive steps by working more closely with state governments to retain farmlands. This was evident in North Carolina, Michigan, Vermont, and New Hampshire. Acting through the Farmers Home Administration, USDA tried to place greater emphasis on rural development and targeted its funding for municipal services away from agricultural areas. By controlling the location of such capital facilities, the Farmers Home Administration expected that the growth-inducing effects of sewer and water investments would not undermine farming.

In 1979 the USDA signed a cooperative agreement with the Council on Environmental Quality to establish and fund a National Agricultural Lands Study to issue a report in 1981 outlining the extent of farmland loss, the costs of such losses, methods to preserve farmlands and enhance agriculture, and a host of other issues related to farm viability. Significantly, federal departments such as Commerce, Defense, Interior, Energy, and Transportation are cosponsoring this study. The Environmental Protection Agency (EPA) finally declared that it intended to preserve farmlands and that all EPA programs and investments would henceforth be evaluated in terms of their impacts upon farmlands.

If the federal role in farmland retention is not direct, this largely results from the traditional American perspective on land use. Land policy is seen as essentially a local and state matter, and those wishing to strengthen the federal role in agricultural land retention have to deal with the broader issue of further federal intrusion into what has historically been a state and local concern.

Some Foreign Examples

The following brief discussions of farmland and agricultural policy in several nations—the Netherlands, Israel, Sweden, the Republic of China (Taiwan), and West Germany—illustrate that other approaches are possible. It is a truism to say that conditions throughout the world vary considerably. In some cases the retention of farmland was necessary because agriculture occupied a strategic and fundamental role in a nation's ideology. In another country, food self-sufficiency was a major goal. And still elsewhere, food as an export commodity was of national significance. What appears obvious from these cases is that the national governments played a vital role in the retention and preservation of farmland and enhancement of agriculture. In some countries this took the form of more effective taxation systems. Other countries developed mechanisms for governmental landownership or land banking. And in still other situations a centralized planning process had sought to protect the agricultural sector from the vagaries of the marketplace. The relevance of these systems to America lies in the fact that while none was totally effective, several worked quite well. Some could be partially replicated depending on the steps localities, states, and the federal government would choose to undertake to address the problem of dwindling farmland resources in the United States.

Most countries with successful programs have in common relatively small size and the fact that they have had bad food scares in the recent past or feared attack by hostile neighbors or both. Taiwan and South Korea were both afraid of enemy attack, being overshadowed by an expansionist China which claimed the Taiwan real estate outright and made vague claims from time to time based on past suzerainty over Korea.

Holland had bitter memories of food shortages during the First and Second World Wars, when some children and adults actually starved, and others suffered the long-run consequences of malnutrition. Also, in the late 1940s, the Netherlands ceased to be an empire upon the loss of most of its colonies. This contributed to a large population increase, from eight million in the 1940s to fourteen million in the 1970s, which mandated increased food production. Holland is a small, compact country with excellent farms, and is bordered by neighbors who had proved hostile in the past. Although for thirty years following World War II, Western Europe had enjoyed peace, the Russian presence in Eastern Europe was continuously disquieting.

Israel is a case apart. Much of the arable land had been reclaimed from desert through some of the most effective programs in the world. Israel had several intervals of war with Arab neighbors, and scarcely a day went by without ominous mutterings. The Iranian revolution of 1979 was especially unsettling, and the Egyptian détente was hopeful but not 100 percent reassuring. Israel was determined to maintain its own food-producing capacity in light of the history of the 1950s, 1960s, and 1970s and in the face of the threats of the 1980s.

America is far different. While America cannot anticipate geographical immunity from a global war, especially one involving rockets and perhaps nuclear weapons, it has never experienced actual hunger, and farmland, considered scarce in the countries mentioned above, was, until the late 1970s, widely considered abundant. The level of motivation to preserve was understandably higher elsewhere.

THE NETHERLANDS

Dutch policy to preserve farmlands from urbanization was part of a larger national planning thrust. Three basic assumptions underlay planning policy in the Netherlands:

1. National and local planners felt the need to make the nation self-sufficient in food production and agriculture.
2. The Dutch sought to control and distribute population among the major cities, the *Randstad* (Rotterdam, the Hague, and Amsterdam), smaller cities, and rural areas. This became necessary as the population expanded to about fourteen million. Given its small land area, the Netherlands was the most densely populated country in the world (990 people per square mile).
3. The nation had long sought to manage its water resources effectively. This meant creating a bulwark against the sea, maintaining and en-

hancing quality fresh waters, and bringing into production new lands reclaimed from the seas—the *polders.*

Farmland retention in the Netherlands included both urban and rural policies. The former sought to consolidate urban lands and halt sprawl, especially between the *Randstad* cities, through a system of land banking. The rural policy consisted, in the main, of consolidation and reparceling of farmland tracts, the public ownership of farmlands in the *polder* area, and a system of subsidies for agricultural production.

The Netherlands has a three-tier system of government—national, provincial, and municipal. The latter includes not only cities but surrounding rural land. Taxes are collected nationally and are allocated to municipalities. National plans must be ratified by the provincial and municipal government, and municipalities retain considerable autonomy. Public participation permeates the Dutch planning process. Many nations had developed national urban planning strategies; the Netherlands went a step further by articulating a national rural policy in the "Report on Rural Areas" and the "Report on Physical Planning in the Netherlands." The concerns of rural areas, and farmlands in particular, were fully integrated into the planning process.

The municipal land banking mechanism established in the late 1960s was essential for the plan and was funded by national tax revenues. It was quite successful in shaping growth, halting sprawl, and preparing the lands to be developed at the right time. Over 80 percent of the land to be developed came from municipal land banks, and over 25 percent of this land was leased rather than sold to private sector developers. The central goal of national planning was to halt urban sprawl in the *Randstad* region, thus preserving farms and open spaces in the most populated area of the country.

When land was purchased by municipalities for future development, compensation was paid for land, structures, and equipment (in the case of farmland) at current use value; there was no uninhibited market to drive up the price of land. All sales and leases from municipal land banks had to be approved at the national level to assure conformance with adopted plans.

National land banking, as opposed to municipal land banking, is predominant only in the *polder* areas—nearly two million acres of land reclaimed from the seas. Most of this region is composed of rich alluvial soils and is ideal for intensive, heavily mechanized farming, especially dairying, grain, and other field-crop production. The large Dutch horticultural industry is not located in the northern and northeastern *polder* region but rather in the agricultural and "greenhouse" areas in and around the *Randstad*.

Farmers from others parts of the Netherlands had been voluntarily relocated in the *polder* areas. Farm sizes in the *polder* areas were substantially larger than in the remainder of the Netherlands, and this permitted greater mechanization. Much of the cost of farm unit rationalization was born by the Land Consolidation Service of the national Agricultural Ministry. Lands were offered for long-term leases, and in an effort to enhance the quality of life in these areas, rural centers and small-size cities were also created to provide diversity, local markets, and social support systems for the farming population.

In summary, Dutch policy to preserve farmland consisted of halting urban sprawl, timing of new development through land banks, the creation of new farmlands (the *polders*), and national ownership and control of such lands. When and where necessary, farmers were relocated at government expense. The Netherlands further aggressively supported its agricultural sector, especially horticulture and bulb-producers, through participation in the European community.

ISRAEL

Though accounting for only 5 percent of Israel's gross national product in the late 1970s, agriculture played a preeminent role in the nation's life and identity. Farming occupied a fundamental role in the ideology of Zionism and thus helped shape the historical development of the nation since the inception of the Jewish National Fund in 1901. Agricultural regions in Israel were utilized as a process for the socialization and integration of immigrants into the fabric of modern Israeli life. Significant water resource planning and development reclaimed hundreds of thousands of acres of arid desert lands. And finally, in a nation almost constantly on a war footing since its establishment in 1948, rural communities formed the basis of a national defense policy and often contributed to the settlement of new regions. The protection and retention of agricultural land had been a critical concern of Israeli policy since the founding of the modern state. Well before World War I, the Jewish National Fund (JNF) began to acquire land for the purpose of establishing agricultural settlements throughout Palestine, although the fund avoided the purchase of lands close to urban areas. Over 30 percent of the present population occupies lands owned by the JNF. With the establishment of modern Israel, it became national policy aggressively to maintain the integrity of the country's farming communities.

Israel's efforts to retain land have been successful for three fundamental reasons:

1. Land use planning was done by the national government and adopted plans were considered legally binding. The lack of a tradition of home rule, together with a long history of immigration and military conflict, enhanced the role of strong central planning and policy making.
2. Over 90 percent of all land in Israel was publicly owned. Private lands were found in the major cities—Tel-Aviv, Haifa, and Jerusalem—and along the coastal plain and were subject to land use controls.
3. Landowners did not have the right to develop land in uses other than those authorized in officially adopted plans, and hence there was little conflict over the kinds of legal issues which were encountered in the United States. A different land use ethic existed in Israel, and when tied to historical themes inherent in Zionism, agricultural lands and farming communities enjoyed significant levels of protection.

Israeli agricultural and rural land policy is epitomized by the example of the Lakhish region, which extends from the coastal plains eastward toward the Judean Hills. Approximately 200,000 acres are within the Lakhish, almost all of it state owned. Though composed of some very rich soils, farm development was not really possible until the construction of the Yarkon-Negev irrigation system. Intensive crop agriculture flourished in the well-irrigated central and western areas. The pasturing of beef and dairy cattle predominated in the less fertile and less irrigated eastern areas, as did small grain production. In all, nearly 45 percent of the Lakhish was farmed. A system of differentiated new communities was created to serve the population of the region, and many are still maturing. A typical area within the Lakhish is the "Nehora Block," which was composed of an urban center, Kiryat Gat; two smaller rural centers, Nehora and Even Shemu'el; several *moshavim* (small holder's cooperative villages), Zohar, Sede Dawid, Kokhav, Uza; and several *kibbutzim* (rural collectives) and *moshavas* (colonies of privately owned and operated small farms). Together these communities and settlements were organized to be mutually supportive in terms of services and amenities. They illustrate the high level of regional cooperation which Israeli planners had sought to achieve.

Israel imports considerable amounts of meat, vegetable oil, and grains, but self-sufficiency has been achieved in fruit and vegetable cultures, poultry, eggs, and dairy products. This sector has the capacity of producing substantial exports, which in the 1970s amounted to some 13 percent of all Israeli exports. This helped a generally negative trade balance. Israel's agricultural economy is quite strong, buttressed by major national investments in training, irrigation, technology, export policy, and based upon a relatively secure farmland base.

SWEDEN

Sweden is a predominantly urban nation with only 10 percent, roughly 7.5 million acres, of its land mass of arable quality. Agricultural production varies with climatic zone change. In the heavily urbanized south, where development pressures were strongest, soils are most favorable to farming. This region produces significant amounts of wheat, sugar beets, peas, and other field crops. In central Sweden the production of bread grains is predominant, and farther north, with a shorter growing season, fodder production assumes a larger role. Perhaps as much as 80 percent of all farm income in Sweden is derived from animal production, with dairying accounting for nearly one-third. ,

Farms in Sweden were decreasing in number, growing in size, and producing more per unit in the 1970s than in previous decades. Swedish agricultural policy in the late 1970s was to encourage these trends, though national planners were seeking to stabilize the number of farming units throughout the nation.

December 1977 was something of a watershed for Swedish agricultural policy. During that month the Swedish Parliament promulgated a number of new programs which have had important impacts upon the structure of the nation's farming industry. National agricultural goals were established to create efficiency of farm size, assure favorable economic returns for farmers, and guarantee the Swedish consumer a reasonable supply of food at reasonable prices.

To achieve these results, the following programs were initiated or continued:

1. Domestic production was protected through tariffs on imported products.
2. Consumer prices were subsidized by the national government.
3. Credit guarantees were made available to farmers to help them expand their units and buy machinery to reach greater economies.

A system of County Agricultural Boards (CABs) was created to aid local farmers within their jurisdictions. Working under the Land Acquisition Act of 1965, the CABs operated directly with local farmers to consolidate local farms into more efficient units. Utilizing monies in the National Land Fund, approximately seventy-three million dollars was spent by 1980 for this purpose. Each purchase had to be approved by the local CAB. Moreover, under other provisions of the law, the CAB monitored the sale of farmland and could refuse to permit a sale if there was reason to believe that the land would be taken out of farm use or be held for speculation. If the CAB refused to allow the sale, it had to purchase the land at the price agreed upon by the farmer and the would-be purchaser. The CAB then held the land until it could sell to a local farmer. Rarely were lands held for a period longer than two years. The loan program developed in the 1977 legislation was one of the key mechanisms to retain Swedish farmland.

Yet some farms were purchased for investment, causing a rise in farmland prices, and a Land Purchase Commission was established to evaluate this problem. It is expected to develop proposals to restrict the purchase of farmlands to those who demonstrate the ability and desire to farm. A strong capital gains tax (*Realisationsvinst*), implemented in 1967, is a further mechanism to discourage speculation.

In the 1970s the average age of the Swedish farmer was fifty-four but increased steadily as young people chose high-paying positions in the country's urban areas. Because it was national policy to sustain family farming, loan guarantees were established to promote the entry of younger farmers. A deferred interest program, with loans payable over a fifteen-year period once farming was fully established, was the key to this program.

As Sweden's population grew, new land was needed for urban development. The CAB system provided for some preemption by urban development. This was especially important in the central and southern regions where growth was sustained, particularly in the suburban areas around such older central cities as Stockholm, Örebro, Gothenburg, and Malmö. Here lands were largely controlled by public land banks operated by urban municipalities. Sweden had a long tradition of municipal planning and local control, and land banking was a major element. The expansion of the metropolitan areas of these cities often meant that farmland was purchased by the municipality to prepare for eventual development. Around Stockholm, 135,000 acres were purchased for such development since the beginning of its land banking system in 1904.

An effort to consolidate urbanization and halt the spread of cities occurred during the 1970s. A location policy for Sweden was established which created twenty-six primary growth centers. It was anticipated that restricting development and growth within these areas would bring a degree of stability to farmland near urban areas, as well as regions where farming predominated.

REPUBLIC OF CHINA (TAIWAN)

The Republic of China on Taiwan was, in the late 1970s, a highly urbanized and industrialized country and another of the most densely populated in the world. With a land mass smaller than the combined areas of Rhode Island, Connecticut, and Massachusetts, Taiwan had over seventeen million people, and over half the land area is mountainous. Yet the nation substantially fed itself and exported considerable amounts of food.

Efforts to stabilize agriculture began with the transfer of the Nationalist government from the mainland to Taiwan in 1949. Following the original precepts of Dr. Sun Yat-sen, founder of modern China, the Nationalists, under General Chiang Kai-shek, initiated a major land reform program which vested land title in the peasants. The peasants purchased this land on terms they could easily manage out of the prior level of farm income.

Taiwan's land reform program consisted of three elements. First, rent control with written leases was established on all privately owned farmlands with rent limited to three-eighths of the rice crop in a given year. Second, lands in the public domain—a large area of arable land which the Japanese had acquired during their fifty years of occupation—were sold to the tillers in 7.5 acre parcels on mortgage terms designed to keep payments also at three-eighths of the annual rice crop. Finally, the central government purchased all rental lands above 7.5 acres and sold them to tillers at equivalent to three-eighths of the rice crop.

The landlords were compensated for their land through bonds, which carried a clause to protect the bondholder in case of inflation, and shares of stock in one of four corporations which had been left by the Japanese. Through this land reform, bona fide small family farmers got land on manageable terms and with it a degree of security which encouraged them to cultivate rice intensively (two annual crops) and also diversify into vegetable production between rice crops. From 1949 to 1955, when the program was substantially completed, the percentage of farmers who owned their lands increased from 57 to 80 percent, while tenant farmers decreased correspondingly. At the same time, efforts by the Joint Commission on Rural Development, a Sino-American agency, encouraged more efficient farming and better conservation. Improvements were made to the rural irrigation system. The result of these programs was a rice yield of approximately one metric ton per acre, one of the highest in the world. The production of bananas, sugar, and vegetables also increased significantly to the point that export markets were developed. Internally the government developed a system of farmers' cooperatives to market produce, supply fertilizer, and extend credit. Rather than creating a new mechanism, the government utilized formerly underground farmers' cooperatives which had grown up during the Japanese occupation. These associations had played a key role during Japanese times as sub rosa bankers, and their credibility served the farmers and the Nationalist government well during the land reform.

In the 1970s a system of land use planning and zoning was established to protect farmlands and to halt urban sprawl. This program was partly effective. As industrialization progressed, sprawl became more and more evident, and farmlands were idled along highways near rapidly growing areas. In 1973 the federal government passed an Agricultural Development Act in an effort to stop the conversion of agricultural lands to other uses. This law required permission of the central agricultural authorities for any conversion of high-grade farmland to other uses. The Regional Planning Act was passed in 1974 and refined the original Chinese Land Act. It remained the principal basis of land use control. It required the classification of all land into one of twenty-six grades and stipulated that:

1. Municipal governments *(Hsien)* designate all farmlands of grades one through twelve as agricultural lands which must be used for farming purposes unless changes were officially authorized.
2. All farmlands above the grade of eight should not be utilized for any construction except farmhouses which must receive prior authorization.
3. Lands between grades nine and twelve should not be converted to in-

dustrial uses without the joint approval of the authorities of industry, agriculture, food, and land administrations.

4. Any deviations from these designations without approval would be regarded as illegal constructions and subject to heavy penalty.

A heavy windfall tax (unearned increment tax) was imposed in the 1940s to take for the government a substantial part of any profits from rising land prices. This incidentally softened the effects of the land control program.

Industrial development expanded throughout Taiwan, and attempts in the late 1970s to direct it onto lands without agricultural potential were relatively successful. The government encouraged the creation of industrial parks and sought to locate them in foothill areas, and it assumed the cost of road construction and also water pollution control. Public facilities, e.g., schools, hospitals, and government offices, were generally developed on public lands, and this further helped the government control the location and timing of development in fringe areas.

A second phase of the Regional Planning Act was implemented in the late 1970s with a pilot program in the southern Taiwan region, containing the cities of Kaohsiung, Tainen, and their rural hinterlands. Under this program, lands were zoned for certain uses, and these would be permanent, single-purpose classifications. The zones to be created included special agriculture, general agriculture, industrial, village, forest, scenic, slope conservation, and special use.

Taiwan also had a most successful program of land consolidation. Rice paddies, for reasons of irrigation, had to be precisely level and have hand-built terraces. The dikes followed contour lines. The result was a patchwork of small fields, and many farmers owned several, sometimes widely scattered. Furthermore, few fields had road frontage, and access was by narrow footpaths along the dikes. The process of consolidation involved the deeding of the small plots to the government, leveling large areas by heavy machine, and reparceling the land back to the former owners, but now in contiguous, rectangular blocks, each with road frontage. Additional benefits were better irrigation and better drainage. The government kept about 10 percent of the land to cover its cost. The value and productivity of the 90 percent returned were better than the small fields given up.

The system worked so well with farmland that it was later extended to city land, and large areas of small holdings were consolidated into lots large enough for apartment projects.

The government of the Republic of Korea (South Korea) implemented a system modeled closely on the successful elements of the Chinese experience on Taiwan.

FEDERAL REPUBLIC OF GERMANY (WEST GERMANY)

Germany has a long history in the control of land use; the framework for all modern land use planning was the Federal Regional Planning Law (BROG) of 1965. Binding upon all states (Länder) and city-states, the BROG established a set of national planning goals. Among these were the creation of a national system of regions with equivalent levels of economic opportunity and social and cultural amenities, the furtherance of regional cooperation among a set of states with relatively high levels of autonomy, and the preservation and protection of the landscape. In more specific terms, the BROG articulated the need to preserve high-quality farmlands, conserve "rural culture," and stem the course of migration from rural areas.

Among the states, Bavaria did the most significant work in agricultural land retention. The Bavarian Land Use Planning Act of 1970 required all political subdivisions to develop comprehensive plans for the equalization of opportunity throughout Bavaria. Fundamental to this was a system of population targets for each of the eighteen subregions in Bavaria keyed to the further development of a set number of growth centers.

The master plan for Bavaria spelled out in very specific terms the need to preserve farms for their cultural as well as obvious economic values. This was especially important in the Alps region where farming was exposed to significant development pressures from recreation and tourism. A system of land use categories, or zones, was established to guide growth and development. These consisted of a development zone, a moderate growth zone, and a nondevelopment zone. The latter included much of the region's agricultural land and, aside from farm buildings and a few dwellings, projected little growth. Government investments, both federal and state, were made in compliance with these use zone categories. Little road construction and few sewage or water systems were funded in farming areas. What growth did occur was targeted for the regional growth centers.

For economic support, a program known as the Mountain Farmer's Subsidy Program was established in 1972 to provide subsidies on a per cow basis to Alpine farmers who committed themselves for a five-year period to farm the high Alpine meadows. Not all farmers subscribed to the program, but it appeared that for those who did it meant the difference between remaining in small-scale agriculture or leaving for a city. This strategy also guaranteed that high-quality landscape and scenic values were maintained, which were essential to Bavaria's huge tourism and recreation industry.

A second program along the same lines was known as "village renewal" which attempted to create more viable village environments which would, in their turn, provide high-level cultural and social amenities to the local farming populations. These areas were designated for some growth and development, though binding limits were enforced. Federal and Bavarian investments helped renew these village centers and upgrade their infrastructures and housing stock. Also, monies were made available to help local farmers consolidate land to realize economies of scale in this dairy region.

Federal legislation in 1949 established a policy of *Flurbereinigung,* land consolidation. State governments might declare that land consolidation was required in certain areas and, with the consent of a majority of local landowners, land exchanges or swaps and rationalizations were carried out. Expropriation was not permitted under German law, and the exchange program attempted to provide each affected landowner with the same total land value—though not necessarily the same acreage—as he had lost. Additionally, any costs for buildings, relocation, or other aspects of the exchange were also paid for by the state government. In those areas where farming had become a marginal economic activity, as in several Alpine towns, land rationalization has been essential.

As in other nations, German policy makers came to understand that beside land use concerns, land tenure and operation issues were fundamental to the viability of agriculture. The Federal Republic is a member of the Common Market and hence participates in the Common Agricultural Policy, which lends further support to agriculture; and, given the nature of land use controls in West Germany, farmers are not under great pressure to convert their farms into other land uses.

Concluding Comments

This discussion has been wide-ranging and, in its brevity, too often superficial. In the United States, most efforts to preserve farmlands are still in their infancy. There has been little federal initiative

178 Mark B. Lapping

in this area, and most state policy has been directed toward land use control. Very few jurisdictions have approached the problem of the disappearance of agricultural land from a comprehensive perspective. Programs which address this and other agriculturally-related issues tend to be highly fragmented, insufficiently defined, and poorly co-ordinated, and focus more on symptoms than on causes and the wider issue of the viability of agriculture in a changing society. Clearly a new federal-state approach is needed, but it must be one which addresses the problems of critical mass, justice, and equity. It must have a flexible concept of what constitutes prime agricultural land, and it must coordinate capital investment with land use goals, structural supports, and incentives for agricultural production. But a more critical issue is how the United States sees itself with regard to agriculture and food production in a world of growing populations and insufficient food supplies.

Index

Abortion, 56
Absentee ownership of rural land, 97, 104-6
Advance acquisition of land for urban development, 71-72
Agricultural Adjustment Administration, 15
Agricultural districting, 161-63
Agricultural price supports, 134
Agricultural research, 79-81, 85-89
Agricultural science, 74-76, 85
Agriculture, Department of (USDA), 32, 36-37, 69, 84, 92, 112, 165
Agrochemicals, 15
Air pollution, 27, 79
American Institute of Architects, 72
American Land Forum, 84
American Law Institute, 72
Animal agriculture, 85
Annapolis, Maryland, 72
Apartment construction, 23
Appalachia, 105
Arbitrage, 130
Automobiles, 27-28, 30-31, 52-53, 61
Aylor, Donald, 4

Baby boom, 14, 48, 52, 57
Back yard gardens, 32-33
Bald eagle, 116
"Balkanized" suburbs, 20
Bavaria, 176-77
Behind Ghetto Walls (Rainwater), 66
Berry, Brian J. L., 36-59, 61, 68
Biological nitrogen fixation, 86, 87-88
Birth control, 56
Birth rate, 14, 38, 56
Blacks, 63-65, 107
Blackstone, Sir William, 18
Bordeaux mixture, 77
Borlaugh, Norman, 81
Buses, 27-28

California, 113, 114, 155, 161, 164
California steppe, 115
Canada, 25
Capital gains tax, 34, 152, 156, 172
Carbon dioxide, 121

Carrying capacity of farmland, 79, 80
Chiang Kai-shek, 173
China, 9
Chrysler Corporation, 52
"Circuit breaker" method, 148
Cities:
  compact, 13, 19
  consumer preferences, 21-22
  counter-urbanization process, 50-53
  densities, 19-21
  "Georgetown" syndrome, 7
  housing market, shaping factors in, 48-50
  industrial exodus from, 24, 25-26
  inner city revitalization, 54-59, 61
  labor force migration, 5-6, 44-48
  mortgage market, 22-24, 28-29, 134
  outer growth and inner decline, 53-54
  population growth and loss, 14, 39-41
  tax, 24-25, 29-30
  transportation, 7, 27-28, 30-31
  vacancies, 6
Classified property tax, 148
Clawson, Marion, 71
Climate, 121
Coastal sagebrush, 115
Commission on Population Growth and the American Future, 2
Common Market Agricultural Policy, 177
Communication, 51-52
Community Block Development Grants, 65
Community land trust, 160-61
Compact cities, 13, 19
Compensable zoning, 154-55
Condominiums, 7, 28-29, 56
Connecticut, 157-58
Connecticut Agricultural Experiment Station, 32
Conservation Foundation, 96
  Rural Land Market Project, 92
Conservation land, 33, 34
Consumer preferences, 21-22
Cooperatives, 29, 56
Corn, 87, 97
Corn Belt, 83, 95, 116, 120, 122
Corporations, 16, 92-93, 97, 105

Corruption, 35
Council on Environmental Quality, 84, 165
Counter-urbanization process, 50-53
Cranberry bogs, 117-18
Credit, 134
Creeping process, 20, 24
Crime rates, 57
Cropland, 111-14
  inventory of, 84
  rate of disappearance of, 3, 11, 82-83
  reserve, 2, 3, 82
  (*see also* Rural landownership)
Crop rotation, 16
Crops, hybrid, 15, 88
Crop surpluses, 15, 18, 19
Crop yields, 15, 86-88

Dark respiration, 86-87
Daylight respiration, 87
Day Waverly Gardens, 32
DDT, 77
Death rate, 38
Decisions for Sale: Corruption and Reform in Land Use and Building Regulation (Gardiner and Lyons), 35
Deferred taxation, 147-51
Deforestation, 117, 121
Denmark, 34
Density, urban, 19-21
Desalination of sea water, 9, 137
Desert land, 9
Development rights purchase, 156-58, 164
Differential assessment, 146-52
Direct marketing, 164
Dirksen, Everett, 69
Discount rates, 130-32
Discrimination, 63
Divorce rate, 39
Douglas Commission, 72
Down-zoning, 10
Drainage, 112, 115, 121
Drug addiction, 67
Dust storms, 16

Ecosystems, endangered, 114-16
Education, 25
Elder, Duncan, 67
Elm-ash forests, 114
Eminent domain, 133, 159
Endangered ecosystems, 114-16
Endangered species, 116
Energy crisis, 5, 18-19, 52-53, 61-62
Environmental Protection Agency (EPA), 52, 165-66
Erosion, 3-4, 15, 16, 103

Evenson, 89
Exclusive agricultural zoning, 153-54
Externalities, 135-37
Exurbia, 61, 62, 70, 72

Family farms, 12, 16, 98-99, 106-7
Family Security Act of 1976, 163-65
Farmers Home Administration, 165
Farming, alternatives to traditional, 8-9
Farmland retention, 34, 69-72, 145-78
  agricultural districting, 161-63
  agricultural viability programs, 163-65
  capital gains tax, 152
  development rights purchase, 156-58, 164
  differential assessment, 146-52
  federal efforts, 165-66
  fee simple purchase and leaseback, 155
  in Israel, 166, 169-71
  land trusts, 160-61
  in Netherlands, 166, 167-69
  in Sweden, 166, 171-73
  in Taiwan, 166, 173-76
  transferable development rights, 158-60
  in West Germany, 166, 176-77
  zoning, 152-55
Federal Housing Administration (FHA), 22-24, 29, 50, 68, 134, 143
Federal Land Banks, 133-34, 143
Federal land policies, 69-72, 124, 131-38, 141, 143-44, 165-66
Federal National Mortgage Association (FNMA), 23-24
Federal Tax Code, 68
Fee simple purchase, 155, 156
Fertility rate, 48
Fertilizers, 8, 15, 75, 111, 121
Feudalism, 143
Fish and Wildlife Service, 116
Fitzharris, 89
Fixed rail systems, 28
Flexible exchange rates, 143
Florida, 114
Food exports, 2, 3, 82, 142-43
Food prices, 78, 94, 142
Food supply, 73, 74, 138
Foreign ownership of land, 16, 93, 95, 97, 105
Foreign trade, 18-19
Forestland, 102-3, 112-14, 119-21
France, 71
Free market (*see* Land market)
Frey, H. Thomas, 37
Fringe farmland, 12, 13, 31-33, 98, 99, 117-19, 134
Frink, Charles R., 1-10, 73-89

Furbish lousewort, 116

Gallen, Hugh, 164
Garden apartments, 23
Gardening, 12, 32-33
Gardiner, J. A., 35
Gasoline, 5, 27-28, 30-31, 52-53, 61
General direction approach, 148
George, Henry, 152
"Georgetown" syndrome, 7
Georgia, 113
Goldschmidt, Walter, 104
Golf courses, 117-18
Government National Mortgage Association (GNMA), 24, 29
Gray, Thomas, 73
Great Britain, 111
Great Depression, 49, 58
Griliches, Zvi, 88, 89

J

Harriss, C. Lowell, 97, 123-44
Hartford, Connecticut, 20, 21
Hatch Act of 1887, 76, 89
Hawaii, 153-54
Healy, Robert G., 90-108
Hedging, 130
Herbicides, 15, 111
Highways, 10, 27, 30, 43, 50, 68, 94
Hobby farm, 100
Homeownership, 49-50, 51, 55-56, 63
Home Owners Loan Corporation, 143
Horsfall, James G., 73-89
Household composition, 39
Housing, 53-54
  blacks and, 63-65
  inner city revitalization, 54-55, 61
  lower income, 62-63, 65-68
  national policy, 49, 51, 53
  shaping factors in market, 48-50
  subsidies, 65-68
Housing Act of 1961, 69
Hunger, 18
Hybrid crops, 15, 88
Hydroponics, 8, 85

Immigration rates, 38
Income tax, 24, 30, 63, 140
Indiana, 114, 115
Industrial exodus from cities, 24, 25-26
Industry, relocation of, 24, 44, 46
Inflation, 52, 55, 62, 78, 94
Inheritance tax, 98-99
Inner city revitalization, 54-59, 61
Inner neighborhood decline, 53-54
Interest rates, 24, 62, 131, 134
Interior, Department of the, 70
Interior, Secretary of the, 70

Internal Revenue Service (IRS), 156
Iowa, 113-15
Irrigation, 78-79, 86, 112
Israel, 9, 10, 137, 166, 169-71

Jewish National Fund (JNF), 169-70
Johnson, Samuel W., 74, 75-76, 80, 81, 88

Labor force migration, 5-6, 44-48
Lake Champlain Islands Trust, 160
Land banking, 71, 133-34, 143
Land-grant college system, 75-76, 81
Landless young people, 97, 106, 163-64
Land market, 123-44
  discount rates, 130-32
  economic and technical flexibility, 137-39
  eminent domain, 133, 159
  externalities, 135-37
  future uncertainties and fixity of supply, 125-26
  fixity and invariability of supply, 126-28
  government and, 124, 131-38, 141, 143-44
  irreversibility, 128-30
  location, 124-25
  mobilizing funds, 133-34
  research, 139-40
  speculation, 129-30
  taxation, 140-41
Landownership (*see* Rural landownership)
Land prices, 95, 100, 101, 106, 127-28, 134
Land shortage:
  global view of, 18-19
  national view of, 15-17
  regional view of, 17-18
Land trusts, 160-61
Lapping, Mark B., 71, 145-78
Leapfrog development, 20, 24, 32, 82, 133, 149
Leaseback, 155
Leased land, 103
Lee, Linda K., 102
Leisure, 51
Liebig, Justus von, 75
Lifestyles, changes in, 55-56, 62
Lockwood Farm, 32
Long-distance shipping, 17, 18
Los Angeles, California, 72
Lower-income housing, 62-63, 65-68
Lyons, T. R., 35

Maine, 164-65
Maine Coastal Heritage Trust, 160

Major Uses of Land in the United States: 1974 (Frey), 37
Malthus, Thomas, 73-74, 76, 79, 89
Manhattan, 72
Manufacturing Belt, 46, 48, 53
Manure, movement of, 21
Marriage rate, 39
Maryland, 146, 158, 159
Massachusetts, 157-58, 165
Meat consumption, 9, 13, 82
Meehan, Eugene, 66
Mekong Valley, 18, 139
Mesquite-live oak savannah, 115
Michigan, 114, 118-19, 165
Middle-class workers, 57
Migrant farm workers, 5-6
Mineral rights, 156
Minimum lot size regulation, 153
Minnesota, 163-64
Minority land ownership, 97, 107
Mississippi, 113
Mississippi River, 138
Mitchell, Parren J., 64
Model Cities, 68
Morrow, Dwight, 134
Mortgage market, 22-24, 28-29, 68, 125, 131, 134, 143
Motives for land purchase, 94-95
Mount Vesuvius, 138

National Academy of Sciences, 2, 3, 79, 80, 81-82, 88
National Agricultural Lands Study, 84
National Association of Conservation Districts, 146
National housing policy, 49, 51, 53
National lands, 70, 119
National Resources Committee, 72
National Water Commission, 2
Netherlands, 10, 34, 71, 166, 167-69
Net photosynthesis, 86-87
Nevada, 114
New Hampshire, 157-58, 164, 165
New Jersey, 146, 157, 159, 164
New York City, 6, 19, 66, 72
New York State, 159, 161-63
Nitrogen fixation, 86, 87-88
North Carolina, 165
Norvell, W. A., 4

Ohio, 113, 114
One-person households, 39
Oregon, 155
Organic food, 8
Our Cities—Their Role in the National Economy (National Resources Committee), 72

Overbuilding, 53-54
Overfarming, 15, 16

Parcel size, 100-103
Parks, 118-20
Pastures, 110, 111, 113, 114
Pennsylvania, 164
Pesticides, 4, 8, 15, 77, 111
Peterson, 89
Phoenix, Arizona, 138
Photorespiration, 87
Photosynthetics, 86-87
Plantation system, 143
Plymouth, Massachusetts, 116
Police power, 154, 159
Pollution, 27, 29
Population, 13, 14, 18, 48, 141
  changes in urban, 39-41
  decline of, 37-38, 62
  percentage changes in, 42
Potato yield, 76-77
Preferential assessment, 147-50
Preservation district, 158
Prices
  food, 78, 94, 142
  land, 95, 100, 101, 106, 127-28, 134
Private land trust, 160
Private market (*see* Land market)
Productivity, 15, 76-77, 79, 81-83, 86-88
Progress and Poverty (George), 152
Property tax, 29-30, 34, 99, 140-41, 148, 151
Protein, 9
Pruit-Igoe project, 66
Public access, 103-4
Public land acquisition systems, 71-72
Public land trust, 160
Public ownership, 71-72

Racial segregation, 63
Rainwater, Lee, 66
Rangeland, 111-14, 120
Rap, John, 71
Recession, 52, 58
Reconstruction Finance Corporation, 143
Recreational land, 33, 34, 94-96, 117-20, 122
Red-bellied turtle, 116
Reforestation, 121
"Regentrification," 64-65
Rent control, 6, 29
Report on Agricultural Production Efficiency (National Academy of Sciences), 80
Republic of China (Taiwan), 9, 10, 166, 173-76

Research:
  agricultural, 79-81, 85-89
  land market, 139-40
Reserve cropland, 2, 3, 82
Restrictive agreement, 147, 150
Retirement, 51, 95
Revitalization of inner cities, 54-55, 61
Rogers, Will, 81
Rollback tax, 147, 149, 150
Rotation of crops, 16
Rowe, Mrs. James, 28
Rural landownership, 15-16, 90-108
  absentee control, 97, 104-6
  consolidating and parceling, 96
  corporate, 16, 92-93, 97, 105
  foreign, 16, 93, 95, 97, 105
  leased land, 103
  market trends, 93-96
  minority, 97, 107
  motives for purchase, 94-95
  ownership and use issues, 96-104
  parcel size, 100-103
  prices, 95, 100, 101, 106, 127-28, 134
  public access, 103-4
  resource control issues, 104
  speculation, 99-100, 129-30, 150, 151, 172
  status, 92-93
Rural landscape, 109-11
Ruttan, 89

Salamander, 116
Salination of land, 3, 15, 78-79, 86
Sand-pine scrub, 114
San Francisco, California, 7, 72
San Francisco Bay National Wildlife Refuge, 116
Santa Fe, New Mexico, 72
Saskatchewan, 155
Savannah, Georgia, 72
Scenic (conservation) easements, 156
Schmitz, 89
Schuh, G. E., 82
Schultz, Theodore W., 85
Sea water, desalination of, 9, 137
Silliman, Benjamin, Jr., 75
Silliman, Benjamin, Sr., 74-75
Singapore, 31
"608 program," 29
Small farms, 12, 16, 98-99, 106-7
Small Farm Viability Planning Project (California), 164
Smith, Frederick E., 109-22
Snail darter, 116
Soil and Water Resources Conservation Act (RCA), 3, 4

Soil banks, 2, 11-12, 78
Soil Conservation Needs Inventory, 82
Soil Conservation Service, 3, 11, 15
Soil erosion, 3-4, 15, 16, 103
Solar energy, 9
Sorghum, 87
South, 43, 48
So. Burlington County NAACP v. Township of Mount Laurel (1972), 153
South Korea, 166
Soviet wheat purchase, 15, 78, 94, 97
Soybeans, 87, 97
Specialization, 126
Species, endangered, 117
Speculation, 99-100, 129-30, 150, 151, 172
Standard Metropolitan Statistical Areas, 17, 21
Stuyvesant, Peter, 1
Suburbia, 61, 62
  "balkanized," 20
  blacks and, 63-64
  industry, relocation of, 24, 25-26
  shopping centers, 24
  zoning, 25, 144
Suffolk County, New York, 157
Sugar cane, 87
Sun Belt, 44
Sun Yat-sen, 173
Surpluses, 15, 18, 19, 77
Sweden, 71, 166, 171-73
Swift, Jonathan, 74, 77

Taiwan, 9, 10, 166, 173-76
Taxation, 24-25, 29-30
  capital gains, 34, 152, 156, 172
  deferral systems, 147-51
  differential assessment, 146-52
  housing investment and, 49-50
  income, 24, 30, 63, 140
  inheritance, 98-99
  property, 29-30, 34, 99, 140-41, 148, 151
  rollback, 147, 149, 150
Tennessee, 164
Texas, 113, 114
Third World, 17, 18
Timberland prices, 95
Timber prices, 94
Time-space convergence, 51
Tobacco and Cotton Belt, 44
Transferable development rights (TDR) proposal, 158-60
Translocation, 87
Transportation, 27-28, 138
  long-distance shipping, 17, 18
  revolution in, 7
  time-space convergence, 51
  urban public, 27-28, 30-31

Tule marsh, 114

Unemployment rate, 57, 58
Upper Great Lakes area, 43
Upper-middle-class intelligentsia, 57
Up-zoning, 10
Urban development (*see* Cities)
Urban fringe, 12, 13, 31-33, 98, 99, 117-
    19, 134
Urbanization, definition of, 12
Utah, 114

Vegetable proteins, 9, 13
Vegetation, 112, 114-15, 121
Vermont, 164, 165
Veterans Administration (VA) loans, 23,
    50
Village of Euclid v. Ambler Realty Co.
    (1929), 152

Waggoner, 89
Wallace, Henry Agard, 78
Wallace, Henry C., 77
Washington, D.C., 7, 28, 72, 115
Washington State, 113, 114
Water erosion, 16
Water shortage, 3, 9, 78
Wealth, concentration of landed, 97, 104

Weaver, Robert C., 60-72
Welfare poor, 56-57
West Germany, 133, 166, 176-77
West Virginia, 164
Wetlands, 8, 111-13, 115, 122
Wheat, exports of, 15, 78, 94, 97
Wilderness, 110, 119
Wildlife refuges, 100, 119, 120
Williamsburg, Virginia, 72
Williamson Act, 149
Wind erosion, 16
Wisconsin, 156
Women, 55-56
Woodruff, A. M., 1-35
Working poor, 56-57
World population, 18
World War II, 49, 50, 58
Wyoming, 113, 114

Youth, landless, 97, 106, 163-64

Zelitch, Israel, 87
Zineb, 77
Zoning, 10, 34, 69, 143, 144, 158
    compensable, 154-55
    exclusive agricultural, 153-54
    minimum lot size regulation, 153
    suburban, 25, 144

The American Assembly
COLUMBIA UNIVERSITY

About The American Assembly

The American Assembly was established by Dwight D. Eisenhower at Columbia University in 1950. It holds nonpartisan meetings and publishes authoritative books to illuminate issues of United States policy.

An affiliate of Columbia, with offices in the Graduate School of Business, the Assembly is a national educational institution incorporated in the State of New York.

The Assembly seeks to provide information, stimulate discussion, and evoke independent conclusions in matters of vital public interest.

AMERICAN ASSEMBLY SESSIONS

At least two national programs are initiated each year. Authorities are retained to write background papers presenting essential data and defining the main issues in each subject.

A group of men and women representing a broad range of experience, competence, and American leadership meet for several days to discuss the Assembly topic and consider alternatives for national policy.

All Assemblies follow the same procedure. The background papers are sent to participants in advance of the Assembly. The Assembly meets in small groups for four or five lengthy periods. All groups use the same agenda. At the close of these informal sessions, participants adopt in plenary session a final report of findings and recommendations.

Regional, state, and local Assemblies are held following the national session at Arden House. Assemblies have also been held in England, Switzerland, Malaysia, Canada, the Caribbean, South America, Central America, the Philippines, and Japan. Over one hundred thirty institutions have co-sponsored one or more Assemblies.

ARDEN HOUSE

Home of the American Assembly and scene of the national sessions is Arden House which was given to Columbia University in 1950 by W. Averell Harriman. E. Roland Harriman joined his brother in contributing toward adaptation of the property for conference purposes. The buildings and surrounding land, known as the Harriman Campus of Columbia University, are 50 miles north of New York City.

Arden House is a distinguished conference center. It is self-supporting and operates throughout the year for use by organizations with educational objectives.